Petroleum Refining Processes

Petroleum Refining Processes

Fidel Russel

Oxford Book Company
Jaipur ▪ New Delhi

Published By

Oxford Book Company

Regd. Off. "Bony Residency", Gate No. 2
Opp. Tilak Public School, Vishweswariya Nagar,
Gopalpura Road, Jaipur - 302018, India
Ph: 0141 - 2761280, Telefax: 2761381
email - oxfordbookjpr@gmail.com
website - oxfordbookcompany.com

Delhi Office: 102, 1st Floor, Satyam House,
4327/3, Ansari Road, Darya Ganj,
New Delhi - 110002, India
Ph: 011-45652440

Distributed by
KEYA INTERNATIONAL PRESS LLP
E-mail: keyainternationalpressllp@gmail.com

First published in India, 2024

ISBN : 9789355240880

Typeset by Shivangi Computers, Jaipur

Printed and bound in India by Replika Press Pvt. Ltd.

Contents

Preface

The reformer–fuel-cell system is still being researched but in the near term, systems would continue to run on existing fuels, such as natural gas or gasoline or diesel. However, there is an active debate about whether using these fuels to make hydrogen is beneficial while global warming is an issue. Fossil fuel reforming does not eliminate carbon dioxide release into the atmosphere but reduces the carbon dioxide emissions as compared to the burning of conventional fuels due to increased efficiency. However, by turning the release of carbon dioxide into a point source rather than distributed release, carbon capture and storage becomes a possibility, which would prevent the carbon dioxide's release to the atmosphere, while adding to the cost of the process. The cost of hydrogen production by reforming fossil fuels depends on the scale at which it is done, the capital cost of the reformer and the efficiency of the unit, so that whilst it may cost only a few dollars per kilogram of hydrogen at industrial scale, it could be more expensive at the smaller scale needed for fuel cells. Recently, a Polish company Bioleux Polska has been advertising renewable hydrogen (RH2) plasma reformers, producing RH2 at under $2 per kilogram and available for lightweight mobile applications using vegetable oil or glycerol as feedstock.

The buffered liquids can be drained to the processing equipment at a much slower rate to prevent overloading the system. As slugs are a periodical phenomenon, the slugcatcher should be emptied before the next slug arrives. Slugcatchers can be used continuously or on-demand. A slugcatcher permanently connected to the pipeline will buffer all production, including the slugs, before it is sent to the gas and liquid handling facilities. This is used for difficult to predict slugging behaviour found in terrain slugging, hydrodynamic slugging or riser-based slugging. Alternatively, the slug catcher can be bypassed in normal operation and be brought online when a slug is expected, usually during pigging operations. An advantage of this set-up is that inspection and maintenance on the slugcatcher can be done without interrupting the normal operation.

Amine gas treating, also known as gas sweetening and acid gas removal, refers to a group of processes that use aqueous solutions of various alkylamines (commonly referred to simply as amines) to remove hydrogen sulfide (H_2S) and carbon dioxide (CO_2) from gases. It is a common unit process used in refineries, and is also used in petrochemical plants, natural gas processing plants and other industries. Processes within oil refineries or chemical processing plants that remove hydrogen sulfide and/or mercaptans are commonly referred to as *sweetening* processes because they result in products which no longer have the sour, foul odors of mercaptans and hydrogen sulfide.

This approach will provide a solid foundation this subject on which the student will continue to build during the remainder of his or her professional life and most certainly during the remainder of his or her professional education.

— ***Fidel Russel***

1

Fossil Fuel Reforming

Fossil fuel reforming is a method of producing hydrogen or other useful products from fossil fuels such as natural gas. This is achieved in a processing device called a reformer which reacts steam at high temperature with the fossil fuel. The steam methane reformer is widely used in industry to make hydrogen. There is also interest in the development of much smaller units based on similar technology to produce hydrogen as a feedstock for fuel cells. Small-scale steam reforming units to supply fuel cells are currently the subject of research and development, typically involving the reforming of methanol or natural gas but other fuels are also being considered such as propane, gasoline, autogas, diesel fuel, and ethanol.

During the conversion of the fossil fuel into hydrogen, carbon is released into the atmosphere, typically as carbon dioxide. As a result, fuel cell systems using reformed fossil fuels would emit substantial amounts of carbon dioxide, so would not make much contribution to reducing carbon dioxide emissions, as is expected to be necessary to tackle global warming.

History

- 1923 - The first synthetic methanol was produced by BASF in Leuna making use of hydrogen derived from lignite.

Industrial Reforming

Steam reforming of natural gas or syngas sometimes referred to as steam methane reforming (SMR) is the most common method of producing commercial bulk hydrogen as well as the hydrogen used in the industrial synthesis of ammonia. It is also the least expensive method. At high temperatures (700 – 1100 °C) and in the presence of a metal-based catalyst (nickel), steam reacts with methane to yield carbon monoxide and hydrogen. These two reactions are reversible in nature.

$$CH_4 + H_2O \rightarrow CO + 3\ H_2$$

Additional hydrogen can be recovered by a lower-temperature gas-shift reaction with the carbon monoxide produced. The reaction is summarized by:

$$CO + H_2O \rightarrow CO_2 + H_2$$

The first reaction is strongly endothermic (consumes heat), the second reaction is mildly exothermic (produces heat).

The United States produces nine million tons of hydrogen per year, mostly with steam reforming of natural gas. The worldwide ammonia production, using hydrogen derived from steam reforming, was 109 million metric tonnes in 2004.

This SMR process is quite different from and not to be confused with catalytic reforming of naphtha, an oil refinery process that also produces significant amounts of hydrogen along with high octane gasoline.

The efficiency of the process is approximately 65% to 75%.

Advantages of Reforming for Supplying Fuel Cells

Steam reforming of gaseous hydrocarbons is seen as a potential way to provide fuel for fuel cells. The basic idea for vehicle on-board reforming is that for example a methanol tank and a steam reforming unit would replace the bulky pressurized hydrogen tanks that would otherwise be necessary. This might mitigate the distribution problems associated with hydrogen vehicles., however the major market players discarded the approach of on-board reforming as unpractical.

Disadvantages of Reforming for Supplying Fuel Cells

The reformer–fuel-cell system is still being researched but in the near term, systems would continue to run on existing fuels, such as natural gas or gasoline or diesel. However, there is an active debate about whether using these fuels to make hydrogen is beneficial while global warming is an issue. Fossil fuel reforming does not eliminate carbon dioxide release into the atmosphere but reduces the carbon dioxide emissions as compared to the burning of conventional fuels due to increased efficiency. However, by turning the release of carbon dioxide into a point source rather than distributed release, carbon capture and storage becomes a possibility, which would prevent the carbon dioxide's release to the atmosphere, while adding to the cost of the process.

The cost of hydrogen production by reforming fossil fuels depends on the scale at which it is done, the capital cost of the reformer and the efficiency of the unit, so that whilst it may cost only a few dollars per kilogram of hydrogen at industrial scale, it could be more expensive at the smaller scale needed for fuel cells. Recently, a Polish company Bioleux Polska has been advertising renewable hydrogen (RH2) plasma reformers, producing RH2 at under $2 per kilogram and available for lightweight mobile applications using vegetable oil or glycerol as feedstock.

Current Problems with Reformers Supplying Fuel Cells

However, there are several challenges associated with this technology:

- The reforming reaction takes place at high temperatures, making it slow to start up and requiring costly high temperature materials.
- Sulfur compounds present in the fuel poison certain catalysts, making it difficult to run this type of system from ordinary gasoline. Some new technologies have overcome this challenge, however, with sulfur-tolerant catalysts.
- Low temperature polymer fuel cell membranes can be poisoned by the carbon monoxide (CO) produced by the reactor, making it necessary to include complex CO-removal systems. Solid oxide fuel cells (SOFC) and Molten carbonate fuel cells (MCFC) do not have this problem, but operate at higher temperatures, slowing start-up time, and requiring costly materials and bulky insulation.
- The thermodynamic efficiency of the process is between 70% and 85% (LHV basis) depending on the purity of the hydrogen product.
- The biggest problem for reformer based systems remains the fuel cell itself, in terms of both cost and durability. The catalyst used in the common polymer-electrolyte-membrane fuel cell, the device most likely to be used in transportation roles, is very sensitive to any leftover carbon monoxide in the fuel, which

some reformers do not completely remove. The anode catalyst is poisoned by the carbon monoxide and the fuel cells performance degrades.

- The catalyst in low temperature fuel cells is based on platinum, and is hence very expensive. A typical automotive fuel cell stack prototype (100 kW) contains 20-30g of platinum metal in the form of nano-particles supported on carbon black.

Slugcatcher

Slugcatcher is the name of a unit in the gas refinery or petroleum industry in which slugs at the outlet of pipelines are collected or caught. A slug is a large quantity of gas or liquid that exits the pipeline.

Slugs

Pipelines that transport both gas and liquids together, known as two-phase flow, can operate in a flow regime known as slugging flow or slug flow. Under the influence of gravity liquids will tend to settle on the bottom of the pipeline, while the gases occupy the top section of the pipeline.

Under certain operating conditions gas and liquid are not evenly distributed throughout the pipeline, but travel as large plugs with mostly liquids or mostly gases through the pipeline. These large plugs are called slugs.

Slugs exiting the pipeline can overload the gas/liquid handling capacity of the plant at the pipeline outlet, as they are often produced at a much larger rate than the equipment is designed for.

Slugs can be generated by different mechanisms in a pipeline:

- Terrain slugging is caused by the elevations in the pipeline, which follows the ground elevation or the sea bed. Liquid can accumulate at a low point of the pipeline until sufficient pressure builds up behind it. Once the liquid is pushed out of the low point, it can form a slug.
- Hydrodynamic slugging is caused by gas flowing at a fast rate over a slower flowing liquid phase. The gas will form waves on the liquid surface, which may grow to bridge the whole cross-section of the line. This creates a blockage on the gas flow, which travels as a slug through the line.

- Riser-based slugging, also known as severe slugging, is associated with the pipeline risers often found in offshore oil production facilities. Liquids accumulate at the bottom of the riser until sufficient pressure is generated behind it to push the liquids over the top of the riser, overcoming the static head. Behind this slug of liquid follows a slug of gas, until sufficient liquids have accumulated at the bottom to form a new liquid slug.
- Pigging slugs are caused by pigging operations in the pipeline. The pig is designed to push all or most of the liquids contents of the pipeline to the outlet. This intentionally creates a liquid slug.

Slugs formed by terrain slugging, hydrodynamic slugging or riser-based slugging are periodical in nature. Whether a slug is able to reach the outlet of the pipeline depends on the rate at which liquids are added to the slug at the front (i.e. in the direction of flow) and the rate at which liquids leave the slug at the back. Some slugs will grow as they travel the pipeline, while others are dampened and disappear before reaching the outlet of the pipeline.

Purpose of the Slugcatcher

A slugcatcher is a vessel with sufficient buffer volume to store the largest slugs expected from the upstream system. The slugcatcher is located between the outlet of the pipeline and the processing equipment.

The buffered liquids can be drained to the processing equipment at a much slower rate to prevent overloading the system. As slugs are a periodical phenomenon, the slugcatcher should be emptied before the next slug arrives.

Slugcatchers can be used continuously or on-demand. A slugcatcher permanently connected to the pipeline will buffer all production, including the slugs, before it is sent to the gas and liquid handling facilities. This is used for difficult to predict slugging behaviour found in terrain slugging, hydrodynamic slugging or riser-based slugging. Alternatively, the slug catcher can be bypassed in normal operation and be brought online when a slug is expected, usually during pigging operations. An advantage of this set-up is that inspection and maintenance on the slugcatcher can be done without interrupting the normal operation.

Slugcatcher Design

Slugcatchers are designed in different forms:

- A vessel type slugcatcher is essentially a conventional vessel. This type is simple in design and maintenance.
- A finger type slugcatcher consists of several long pieces of pipe ('fingers'), which together form the buffer volume. The advantage of this type of slugcatcher is that pipe segments are simpler to design for high pressures, which are often encountered in pipeline systems, than a large vessel. A disadvantage is that its footprint can become excessively large. An example of a large finger-type slugcatcher can be seen in Den Helder, The Netherlands, using Google Maps.
- A Parking Loop slugcatcher combines features of the vessel and finger types. The Gas/Liquid Separation occurs in the Vessel, while the Liquid is stored in the parking loop shaped fingers.

A basic slug catcher design contains the buffer volume for gas and liquid. A control system is used for controlled outflow of gas and liquid to the downstream processing facilities. The inlet section is designed to promote the separation of gas and liquid.

Amine Gas Treating

Amine gas treating, also known as gas sweetening and acid gas removal, refers to a group of processes that use aqueous solutions of various alkylamines (commonly referred to simply as amines) to remove hydrogen sulfide (H_2S) and carbon dioxide (CO_2) from gases. It is a common unit process used in refineries, and is also used in petrochemical plants, natural gas processing plants and other industries.

Processes within oil refineries or chemical processing plants that remove hydrogen sulfide and/or mercaptans are commonly referred to as *sweetening* processes because they result in products which no longer have the sour, foul odors of mercaptans and hydrogen sulfide.

There are many different amines used in gas treating:

- Monoethanolamine (MEA)
- Diethanolamine (DEA)
- Methyldiethanolamine (MDEA)

- Diisopropylamine (DIPA)
- Aminoethoxyethanol (diglycolamine) (DGA).

The most commonly used amines in industrial plants are the alkanolamines MEA, DEA, and MDEA.

Amines are also used in many oil refineries to remove sour gases from liquid hydrocarbons such as liquified petroleum gas (LPG).

Description of a Typical Amine Treater

Gases containing H_2S or both H_2S and CO_2 are commonly referred to as *sour gases* or *acid gases* in the hydrocarbon processing industries.

The chemistry involved in the amine treating of such gases varies somewhat with the particular amine being used. For one of the more common amines, methanolamine (MEA) denoted as RNH_2, the chemistry may be simply expressed as:

$$RNH_2 + H_2S \qquad RNH_3HS$$

A typical amine gas treating process (as shown in the flow diagram below) includes an absorber unit and a regenerator unit as well as accessory equipment. In the absorber, the downflowing amine solution absorbs H_2S and CO_2 from the upflowing sour gas to produce a sweetened gas stream (i.e., an H_2S-free gas) as a product and an amine solution rich in the absorbed acid gases.

The resultant "rich" amine is then routed into the regenerator (a stripper with a reboiler) to produce regenerated or "lean" amine that is recycled for reuse in the absorber. The stripped overhead gas from the regenerator is concentrated H_2S and CO_2. In oil refineries, that stripped gas is mostly H_2S, much of which often comes from a sulfur-removing process called hydrodesulfurization.

This H_2S-rich stripped gas stream is then usually routed into a Claus process to convert it into elemental sulfur. In fact, the vast majority of the 64,000,000 metric tons of sulfur produced worldwide in 2005 was byproduct sulfur from refineries and other hydrocarbon processing plants. Another sulfur-removing process is the WSA Process which recovers sulfur in any form as concentrated sulfuric acid. In some plants, more than one amine absorber unit may share a common regenerator unit.

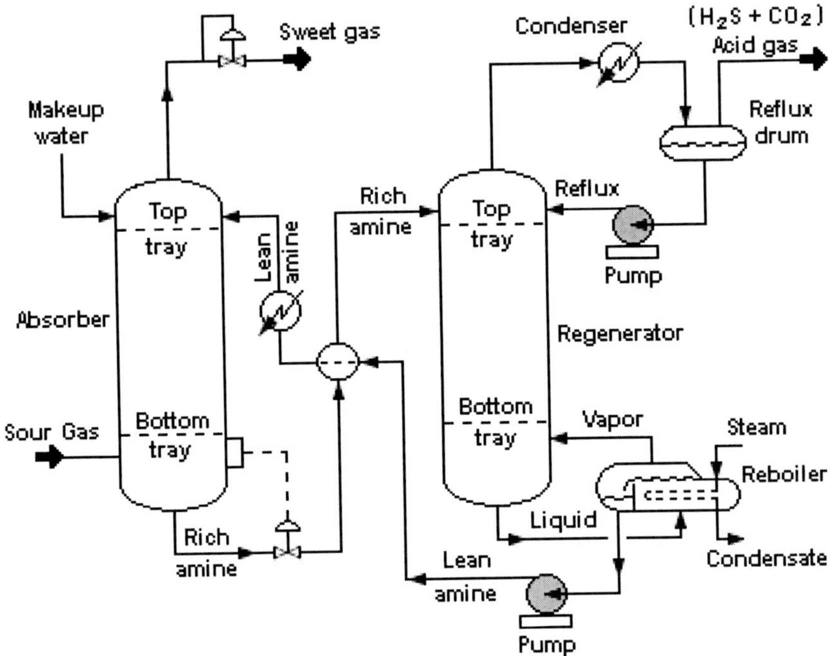

Typical operating ranges

Absorber : 35 to 50 °C and 5 to 205 atm of absolute pressure
Regenerator : 115 to 126 °C and 1.4 to 1.7 atm of absolute pressure at tower bottom

Figure 1 : Process flow diagram of a typical amine treating process used in industrial plants.

The amine concentration in the absorbent aqueous solution is an important parameter in the design and operation of an amine gas treating process. Depending on which one of the following four amines the unit was designed to use and what gases it was designed to remove, these are some typical amine concentrations, expressed as weight percent of pure amine in the aqueous solution:

- Monoethanolamine: About 20 % for removing H_2S and CO_2, and about 32 % for removing only CO_2.
- Diethanolamine: About 20 to 25 % for removing H_2S and CO_2
- Methyldiethanolamine: About 30 to 55% % for removing H_2S and CO_2
- Diglycolamine: About 50 % for removing H_2S and CO_2

The choice of amine concentration in the circulating aqueous solution depends upon a number of factors and may be quite arbitrary.

It is usually made simply on the basis of experience. The factors involved include whether the amine unit is treating raw natural gas or petroleum refinery by-product gases that contain relatively low concentrations of both H_2S and CO_2 or whether the unit is treating gases with a very high percentage of CO_2 such as the offgas from the steam reforming process used in ammonia production or the flue gases from power plants.

Both H_2S and CO_2 are acid gases and hence corrosive to carbon steel. However, in an amine treating unit, CO_2 is the stronger acid of the two. H_2S forms a film of iron sulfide on the surface of the steel that acts to protect the steel. When treating gases with a very high percentage of CO_2, corrosion inhibitors are often used and that permits the use of higher concentrations of amine in the circulating solution. Another factor involved in choosing an amine concentration is the relative solubility of H_2S and CO_2 in the selected amine.

The choice of the type of amine will affect the required circulation rate of amine solution, the energy consumption for the regeneration and the ability to selectively remove either H_2S alone or CO_2 alone if desired.

The current emphasis on removing CO_2 from the flue gases emitted by fossil fuel power plants has led to much interest in using amines for that purpose.

In the specific case of the industrial synthesis of ammonia, for the steam reforming process of hydrocarbons to produce gaseous hydrogen, amine treating is one of the commonly used processes for removing excess carbon dioxide in the final purification of the gaseous hydrogen.

Claus Process

The Claus process is the most significant gas desulfurizing process, recovering elemental sulfur from gaseous hydrogen sulfide. First patented in 1883 by the scientist *Carl Friedrich Claus*, the Claus process has become the industry standard.

The multi-step Claus process recovers sulfur from the gaseous hydrogen sulfide found in raw natural gas and from the by-product gases containing hydrogen sulfide derived from refining crude oil and other industrial processes. The by-product gases mainly originate from physical and chemical gas treatment units (Selexol, Rectisol,

Purisol and amine scrubbers) in refineries, natural gas processing plants and gasification or synthesis gas plants. These by-product gases may also contain hydrogen cyanide, hydrocarbons, sulfur dioxide or ammonia.

Gases with an H_2S content of over 25% are suitable for the recovery of sulfur in straight-through Claus plants while alternate configurations such as a split-flow set up or feed and air preheating can be used to process leaner feeds.

Hydrogen sulfide produced, for example, in the hydro-desulfurization of refinery naphthas and other petroleum oils, is converted to sulfur in Claus plants. The overall main reaction equation is:

$$2\ H_2S + O_2 \rightarrow S_2 + 2\ H_2O$$

In fact, the vast majority of the 64,000,000 metric tons of sulfur produced worldwide in 2005 was byproduct sulfur from refineries and other hydrocarbon processing plants. Sulfur is used for manufacturing sulfuric acid, medicine, cosmetics, fertilizers and rubber products. Elemental sulfur is used as fertilizer and pesticide.

History

The process was invented by Carl Friedrich Claus, a chemist working in England. A British patent was issued to him in 1883. The process was later significantly modified by a German company called IG Farben.

Thermal Step

In the thermal step, hydrogen sulfide-laden gas reacts in a substoichiometric combustion at temperatures above 850 °C such that elemental sulfur precipitates in the downstream process gas cooler.

The H_2S content and the concentration of other combustible components (hydrocarbons or ammonia) determine the location where the feed gas is burned. Claus gases (acid gas) with no further combustible contents apart from H_2S are burned in lances surrounding a central muffle by the following chemical reaction:

$$2\ H_2S + 3\ O_2 \rightarrow 2\ SO_2 + 2\ H_2O \qquad (\Delta H = -4147.2\ \text{kJ mol}^{-1})$$

This is a strongly exothermic free-flame total oxidation of hydrogen sulfide generating sulfur dioxide that reacts away in subsequent reactions. The most important one is the Claus reaction:

$2\ H_2S + SO_2 \rightarrow 3\ S + 2\ H_2O$

The overall equation is :

$10\ H_2S + 5\ O_2 \rightarrow 2\ H_2S + SO_2 + 7/2\ S_2 + 8\ H_2O$

This equation shows that in the thermal step alone two-thirds of the hydrogen sulfide is converted to sulfur.

Gases containing ammonia, such as the gas from the refinery's sour water stripper (SWS), or hydrocarbons are converted in the burner muffle. Sufficient air is injected into the muffle for the complete combustion of all hydrocarbons and ammonia. The air to the acid gas ratio is controlled such that in total 1/3 of all hydrogen sulfide (H_2S) is converted to SO_2. This ensures a stoichiometric reaction for the Claus reaction in the second catalytic step.

The separation of the combustion processes ensures an accurate dosage of the required air volume needed as a function of the feed gas composition. To reduce the process gas volume or obtain higher combustion temperatures, the air requirement can also be covered by injecting pure oxygen. Several technologies utilizing high-level and low-level oxygen enrichment are available in industry, which requires the use of a special burner in the reaction furnace for this process option.

Usually, 60 to 70% of the total amount of elemental sulfur produced in the process are obtained in the thermal process step.

The main portion of the hot gas from the combustion chamber flows through the tube of the process gas cooler and is cooled down such that the sulfur formed in the reaction step condenses. The heat given off by the process gas and the condensation heat evolved are utilized to produce medium or low-pressure steam. The condensed sulfur is removed at the gas outlet section of the process gas cooler.

A small portion of the process gas can be routed through a bypass inside of the process gas cooler, as depicted in the here above mentioned figure. This hot bypass stream is added to the cold process gas through a three-way valve to adjust the inlet temperature required for the first reactor.

The sulfur forms in the thermal phase as highly reactive S_2 diradicals which combine exclusively to the S_8 allotrope:

$4\ S_2 \rightarrow S_8$

Side Reactions

Other chemical processes taking place in the thermal step of the Claus reaction are :

- The formation of hydrogen gas:

$2\ H_2S \rightarrow S_2 + 2\ H_2 \qquad (\Delta H > 0)$

$CH_4 + 2\ H_2O \rightarrow CO_2 + 4\ H_2$

- The formation of carbonyl sulfide:

$H_2S + CO_2 \rightarrow S{=}C{=}O + H_2O$

- The formation of carbon sulfide:

$CH_4 + 2\ S_2 \rightarrow S{=}C{=}S + 2\ H_2S$

Catalytic Step

The Claus reaction continues in the catalytic step with activated aluminum(III) or titanium(IV) oxide, and serves to boost the sulfur yield. More hydrogen sulfide (H_2S) reacts with the SO_2 formed during combustion in the reaction furnace in the Claus reaction, and results in gaseous, elemental sulfur.

$2\ H_2S + SO_2 \rightarrow 3\ S + 2\ H_2O \qquad (\Delta H = -1165.6\ \text{kJ mol}^{-1})$

This sulfur can be S_6, S_7, S_8 or S_9.

The catalytic recovery of sulfur consists of three substeps: heating, catalytic reaction and cooling plus condensation. These three steps are normally repeated a maximum of three times. Where an incineration or tail-gas treatment unit (TGTU) is added downstream of the Claus plant, only two catalytic stages are usually installed.

The first process step in the catalytic stage is the gas heating process. It is necessary to prevent sulfur condensation in the catalyst bed, which can lead to catalyst fouling. The required bed operating temperature in the individual catalytic stages is achieved by heating the process gas in a reheater until the desired operating bed temperature is reached.

Several methods of reheating are used in industry:

- Hot-gas bypass: which involves mixing the two process gas streams from the process gas cooler (cold gas) and the bypass (hot gas) from the first pass of the waste-heat boiler.
- Indirect steam reheaters: the gas can also be heated with high-pressure steam in a heat exchanger.

- Gas/gas exchangers: whereby the cooled gas from the process gas cooler is indirectly heated from the hot gas coming out of an upstream catalytic reactor in a gas-to-gas exchanger.
- Direct-fired heaters: fired reheaters utilizing acid gas or fuel gas, which is burned substoichiometrically to avoid oxygen breakthrough which can damage Claus catalyst.

The typically recommended operating temperature of the first catalyst stage is 315 °C to 330 °C (bottom bed temperature). The high temperature in the first stage also helps to hydrolyze COS and CS_2, which is formed in the furnace and would not otherwise be converted in the modified Claus process.

The catalytic conversion is maximized at lower temperatures, but care must be taken to ensure that each bed is operated above the dew point of sulfur. The operating temperatures of the subsequent catalytic stages are typically 240 °C for the second stage and 200 °C for the third stage (bottom bed temperatures).

In the sulfur condenser, the process gas coming from the catalytic reactor is cooled to between 150 and 130 °C. The condensation heat is used to generate steam at the shell side of the condenser.

Before storage, liquid sulfur streams from the process gas cooler, the sulfur condensers and from the final sulfur separator are routed to the degassing unit, where the gases (primarily H_2S) dissolved in the sulfur are removed.

The tail gas from the Claus process still containing combustible components and sulfur compounds (H_2S, H_2 and CO) is either burned in an incineration unit or further desulfurized in a downstream tail gas treatment unit.

Sub Dew Point Claus Process

The conventional Claus process described above is limited in its conversion due to the reaction equilibrium being reached. Like all exothermic reactions, greater conversion can be achieved at lower temperatures, however as mentioned the Claus reactor must be operated above the sulfur dew point (120-150 °C) to avoid liquid sulfur physically deactivating the catalyst. To overcome this problem, the sub dew point Claus process operates with reactors in parallel. When one reactor has become saturated with adsorbed sulfur, the process flow is diverted to the standby reactor. The reactor is then regenerated by sending process gas that has been heated to 300-350 °C to vaporize the sulfur. This stream is sent to a condenser to recover the sulfur.

Process Performance

Using two catalytic stages, the process will typically yield over 97% of the sulfur in the input stream. Over 2.6 tons of steam will be generated for each ton of sulfur yield.

The physical properties of elemental sulfur obtained in the Claus process can differ from that obtained by other processes. Sulfur is usually transported as a liquid (melting point 115 °C). In ordinary sulfur viscosity can increase rapidly at temperatures in excess of 160 °C due to the formation of polymeric sulfur chains but not so in Claus-sulfur. Another anomaly is found in the solubility of residual H_2S in liquid sulfur as a function of temperature. Ordinarily the solubility of a gas decreases with increasing temperature but now it is the opposite. This means that toxic and explosive H_2S gas can build up in the headspace of any cooling liquid sulfur reservoir. The explanation for this anomaly is the endothermic reaction of sulfur with H_2S to polysulfane.

Hydrogen Sulfide

Hydrogen sulfide (British English: *hydrogen sulphide*) is the chemical compound with the formula H_2S. It is a colorless, very poisonous, flammable gas with the characteristic foul odor of expired eggs perceptible at concentrations as low as 0.00047 parts per million. It often results from the bacterial breakdown of organic matter in the absence of oxygen, such as in swamps and sewers; this process is commonly known as anaerobic digestion. It also occurs in volcanic gases, natural gas, and some well waters. The human body produces small amounts of H_2S and uses it as a signaling molecule.

Properties

Hydrogen sulfide is slightly heavier than air; a mixture of H_2S and air is explosive. Hydrogen sulfide and oxygen burn with a blue flame to form sulfur dioxide (SO_2) and water. In general, hydrogen sulfide acts as a reducing agent.

At high temperature or in the presence of catalysts, sulfur dioxide can be made to react with hydrogen sulfide to form elemental sulfur and water. This is exploited in the Claus process, the main way to convert hydrogen sulfide into elemental sulfur.

Hydrogen sulfide is slightly soluble in water and acts as a weak acid, giving the hydrosulfide ion HS^- (pK_a = 6.9 in 0.01-0.1 mol/litre solutions at 18 °C) and the sulfide ion S^{2-} (pK_a = 11.96).

A solution of hydrogen sulfide in water, known as *sulfhydric acid* or *hydrosulfuric acid*, is initially clear but over time turns cloudy. This is due to the slow reaction of hydrogen sulfide with the oxygen dissolved in water, yielding elemental sulfur, which precipitates out.

Hydrogen sulfide reacts with metal ions to form metal sulfides, which may be considered the salts of hydrogen sulfide. Some ores are sulfides. Metal sulfides often have a dark colour.

Lead(II) acetate paper is used to detect hydrogen sulfide because it turns grey in the presence of the gas as lead(II) sulfide is produced. Reacting metal sulfides with strong acid liberates hydrogen sulfide.

If gaseous hydrogen sulfide is put into contact with concentrated nitric acid, it explodes.

Hydrogen sulfide reacts with alcohols to form thiols.

Production

Hydrogen sulfide is most commonly obtained by its separation from sour gas, which is natural gas with high content of H_2S. It can also be produced by reacting hydrogen gas with molten elemental sulfur at about 450 °C. Hydrocarbons can replace hydrogen in this process.

Sulfate-reducing bacteria (resp. sulfur-reducing bacteria) generate usable energy under low-oxygen conditions by using sulfates (resp. elemental sulfur) to oxidize organic compounds or hydrogen; this produces hydrogen sulfide as a waste product.

The standard lab preparation is to react ferrous sulfide (FeS) with a strong acid in a Kipp generator.

$$FeS + 2\ HCl \rightarrow FeCl_2 + H_2S$$

A less well-known and more convenient alternative is to react aluminium sulfide with water:

$$6\ H_2O + Al_2S_3 \rightarrow 3\ H_2S + 2\ Al(OH)_3$$

This gas is also produced by heating sulfur with solid organic compounds and by reducing sulfurated organic compounds with hydrogen.

Hydrogen sulfide is also a byproduct of some reactions and caution should be taken when production is likely as exposure can be fatal.

Occurrence

Small amounts of hydrogen sulfide occur in crude petroleum, but natural gas can contain up to 90%. Volcanoes and some hot springs (as well as cold springs) emit some H_2S, where it probably arises via the hydrolysis of sulfide minerals, i.e. $MS + H_2O \rightarrow MO + H_2S$.

About 10% of total global emissions of H_2S is due to human activity. By far the largest industrial route to H_2S occurs in petroleum refineries: The hydrodesulfurization process liberates sulfur from petroleum by the action of hydrogen.

The resulting H_2S is converted to elemental sulfur by partial combustion via the Claus process, which is a major source of elemental sulfur.

Other anthropogenic sources of hydrogen sulfide include coke ovens, paper mills (using the sulfate method), and tanneries. H_2S arises from virtually anywhere where elemental sulfur comes in contact with organic material, especially at high temperatures.

Hydrogen sulfide can be present naturally in well water. In such cases, ozone is often used for its removal; an alternative method uses a filter with manganese dioxide. Both methods oxidize sulfides to much less toxic sulfates.

Uses

Production of Thioorganic Compounds

Several organosulfur compounds are produced using hydrogen sulfide. These include methanethiol, ethanethiol, and thioglycolic acid.

Alkali Metal Sulfides

Upon combining with alkali metal bases, hydrogen sulfide converts to alkali hydrosulfides such as sodium hydrosulfide and sodium sulfide, which are used in the degradation of biopolymers. The depilation of hides and the delignification of pulp by the Kraft process both are effected by alkali sulfides.

Analytical Chemistry

For well over a century, hydrogen sulfide was important in analytical chemistry, in the qualitative inorganic analysis of metal ions. In these analyses, heavy metal (and nonmetal) ions (e.g., Pb(II), Cu(II), Hg(II), As(III)) are precipitated from solution upon exposure to H_2S. The components of the resulting precipitate redissolve with some selectivity.

For small-scale laboratory use in analytic chemistry, the use of thioacetamide has superseded H_2S as a source of sulfide ions.

Precursor to Metal Sulfides

As indicated above, many metal ions react with hydrogen sulfide to give the corresponding metal sulfides. This conversion is widely exploited. For example, gases or waters contaminated by hydrogen sulfide can be cleaned with metal sulfides. In the purification of metal ores by flotation, mineral powders are often treated with hydrogen sulfide to enhance the separation.

Metal parts are sometimes passivated with hydrogen sulfide. Catalysts used in hydrodesulfurization are routinely activated with hydrogen sulfide, and the behaviour of metallic catalysts used in other parts of a refinery is also modified using hydrogen sulfide.

Miscellaneous Applications

Hydrogen sulfide is used to separate deuterium oxide, or heavy water, from normal water via the Girdler Sulfide process.

Removal from Fuel Gases

Hydrogen sulfide is commonly found in natural gas, biogas, and LPG. It can be removed in a number of ways.

Reaction with Iron Oxide

Gas is pumped through a container of hydrated iron(III) oxide, which combines with hydrogen sulfide.

$$Fe_2O_3(s) + H_2O(l) + 3\ H_2S(g) \rightarrow Fe_2S_3(s) + 4\ H_2O(l)$$

In order to regenerate iron(III) oxide, the container must be taken out of service, flooded with water and aerated.

$$2\ Fe_2S_3(s) + 3\ O_2(g) + 2\ H_2O(l) \rightarrow 2\ Fe_2O_3(s) + 2H_2O(l) + 6\ S(s)$$

On completion of the regeneration reaction the container is drained of water and can be returned to service.

The advantage of this system is that it is completely passive during the extraction phase.

Hydrodesulfurization

Hydrodesulfurization is a more complex method of removing sulfur from fuels.

Safety

Hydrogen sulfide is a highly toxic and flammable gas (flammable range: 4.3–46%). Being heavier than air, it tends to accumulate at the bottom of poorly ventilated spaces. Although very pungent at first, it quickly deadens the sense of smell, so potential victims may be unaware of its presence until it is too late. For safe handling procedures, a hydrogen sulfide material safety data sheet (MSDS) should be consulted.

In 1975, a hydrogen sulfide explosion in Denver City, located in Yoakum and Gaines counties, Texas, caused the state legislature to focus on the deadly hazards of the gas. State Representative E L Short of Tahoka in Lynn County, took the lead in endorsing an investigation by the Texas Railroad Commission and urged that residents be warned "by knocking on doors if necessary" of the imminent danger stemming from the gas. One may die from the second inhalation of the gas, and a warning itself may be too late.

Toxicity

Hydrogen sulfide is considered a broad-spectrum poison, meaning that it can poison several different systems in the body, although the nervous system is most affected. The toxicity of H_2S is comparable with that of hydrogen cyanide. It forms a complex bond with iron in the mitochondrial cytochrome enzymes, thus preventing cellular respiration.

Since hydrogen sulfide occurs naturally in the body, the environment and the gut, enzymes exist in the body capable of detoxifying it by oxidation to (harmless) sulfate. Hence, low levels of hydrogen sulfide may be tolerated indefinitely.

At some threshold level, believed to average around 300–350 ppm, the oxidative enzymes become overwhelmed. Many personal safety gas detectors, such as those used by utility, sewage and petrochemical workers, are set to alarm at as low as 5 to 10 ppm and to go into high alarm at 15 ppm.

An interesting diagnostic clue of extreme poisoning by H_2S is the discoloration of copper coins in the pockets of the victim. Treatment involves immediate inhalation of amyl nitrite, injections of sodium nitrite, inhalation of pure oxygen, administration of bronchodilators to overcome eventual bronchospasm, and in some cases hyperbaric oxygen therapy (HBO). HBO therapy has anecdotal support and remains controversial.

Exposure to lower concentrations can result in eye irritation, a sore throat and cough, nausea, shortness of breath, and fluid in the lungs. These effects are believed to be due to the fact that hydrogen sulfide combines with alkali present in moist surface tissues to form sodium sulfide, a caustic. These symptoms usually go away in a few weeks.

Long-term, low-level exposure may result in fatigue, loss of appetite, headaches, irritability, poor memory, and dizziness. Chronic exposure to low level H_2S (around 2 ppm) has been implicated in increased miscarriage and reproductive health issues among Russian and Finnish wood pulp workers, but the reports have not (as of circa 1995) been replicated.

- 0.00047 ppm is the recognition threshold, the concentration at which 50% of humans can detect the characteristic odor of hydrogen sulfide, normally described as resembling "a rotten egg".
- Less than 10 ppm has an exposure limit of 8 hours per day.
- 10–20 ppm is the borderline concentration for eye irritation.
- 50–100 ppm leads to eye damage.
- At 100–150 ppm the olfactory nerve is paralyzed after a few inhalations, and the sense of smell disappears, often together with awareness of danger.
- 320–530 ppm leads to pulmonary edema with the possibility of death.
- 530–1000 ppm causes strong stimulation of the central nervous system and rapid breathing, leading to loss of breathing.
- 800 ppm is the lethal concentration for 50% of humans for 5 minutes exposure (LC50).
- Concentrations over 1000 ppm cause immediate collapse with loss of breathing, even after inhalation of a single breath.

Although respiratory paralysis may be immediate, it can also be delayed up to 72 hours.

Hydrogen sulfide was used by the British as a chemical agent during World War I. It was not considered to be an ideal war gas, but, while other gases were in short supply, it was used on two occasions in 1916.

A dump of toxic waste containing hydrogen sulfide is believed to have caused 17 deaths and thousands of illnesses in Abidjan, on the West Africa coast, in the 2006 Côte d'Ivoire toxic waste dump.

Suicides

The gas, produced by mixing certain household ingredients, was used in a suicide wave in 2008 in Japan. The wave prompted staff at Tokyo's suicide prevention centre to set up a special hot line during "Golden Week", as they received an increase in calls from people wanting to kill themselves during the annual May holiday.

As of 2010, this phenomenon has occurred in a number of US cities (and in Putney West London, England), prompting warnings to those arriving at the site of the suicide. These first responders, such as emergency services workers or family members are at risk of death from inhaling lethal quantities of the gas, or by fire. Local governments have also initiated campaigns to prevent such suicides.

Function in the Body

Hydrogen sulfide is produced in small amounts by some cells of the mammalian body and has a number of biological signaling functions. (Only two other such gases are currently known: nitric oxide (NO) and carbon monoxide (CO).)

The gas is produced from cysteine by the enzymes cystathionine beta-synthase and cystathionine gamma-lyase. It acts as a relaxant of smooth muscle and as a vasodilator and is also active in the brain, where it increases the response of the NMDA receptor and facilitates long term potentiation, which is involved in the formation of memory.

Eventually the gas is converted to sulfite in the mitochondria by thiosulfate reductase, and the sulfite is further oxidized to thiosulfate and sulfate by sulfite oxidase. The sulfates are excreted in the urine.

Due to its effects similar to nitric oxide (without its potential to form peroxides by interacting with superoxide), hydrogen sulfide is now recognized as potentially protecting against cardiovascular disease.

The cardioprotective role effect of garlic is caused by catabolism of the polysulfide group in allicin to H_2S, a reaction that could depend on reduction mediated by glutathione.

Though both nitric oxide and hydrogen sulfide have been shown to relax blood vessels, their mechanisms of action are different: while NO activates the enzyme guanylyl cyclase, H_2S activates ATP-sensitive potassium channel in smooth muscle cells. Researchers are not clear how the vessel-relaxing responsibilities are shared between nitric oxide and hydrogen sulfide. However there exists some evidence to suggest that nitric oxide does most of the vessel-relaxing work in large vessels and hydrogen sulfide is responsible for similar action in smaller blood vessels.

Like nitric oxide, hydrogen sulfide is involved in the relaxation of smooth muscle that causes erection of the penis, presenting possible new therapy opportunities for erectile dysfunction.

In Alzheimer's disease the brain's hydrogen sulfide concentration is severely decreased. In trisomy 21 (the most common form of Down syndrome) the body produces an excess of hydrogen sulfide. Hydrogen sulfide is also involved in the disease process of type 1 diabetes. The beta cells of the pancreas in type 1 diabetes produce an excess of the gas, leading to the death of beta cells and to a reduced production of insulin by those that remain.

Induced Hypothermia/Suspended Animation

In 2005, it was shown that mice can be put into a state of suspended animation-like hypothermia by applying a low dosage of hydrogen sulfide (81 ppm H_2S) in the air. The breathing rate of the animals sank from 120 to 10 breaths per minute and their temperature fell from 37 °C to just 2 °C above ambient temperature (in effect, they had become cold-blooded). The mice survived this procedure for 6 hours and afterwards showed no negative health consequences. In 2006 it was shown that the blood pressure of mice treated in this fashion with hydrogen sulfide did not significantly decrease.

A similar process known as hibernation occurs naturally in many mammals and also in toads, but not in mice. (Mice can fall into a state called clinical torpor when food shortage occurs). If the H_2S-induced hibernation can be made to work in humans, it could be useful in the emergency management of severely injured patients, and in the conservation of donated organs. In 2008, hypothermia induced by hydrogen sulfide for 48 hours was shown to reduce the extent of brain damage caused by experimental stroke in rats.

As mentioned above, hydrogen sulfide binds to cytochrome oxidase and thereby prevents oxygen from binding, which leads to the dramatic slowdown of metabolism. Animals and humans naturally produce some hydrogen sulfide in their body; researchers have proposed that the gas is used to regulate metabolic activity and body temperature, which would explain the above findings.

Two recent studies cast doubt that the effect can be achieved in larger mammals. A 2008 study failed to reproduce the effect in pigs, concluding that the effects seen in mice were not present in larger mammals. Likewise a paper by Haouzi et al. noted that there is no induction of hypometabolism in sheep, either.

However, at a February 2010 TED conference, Mark Roth announced that hydrogen sulfide induced hypothermia had completed Phase I clinical trials. He estimated that further trials would take 'a few years.'

Participant in the Sulfur Cycle

Hydrogen sulfide is a central participant in the sulfur cycle, the biogeochemical cycle of sulfur on Earth.

In the absence of oxygen, sulfur-reducing and sulfate-reducing bacteria derive energy from oxidizing hydrogen or organic molecules by reducing elemental sulfur or sulfate to hydrogen sulfide. Other bacteria liberate hydrogen sulfide from sulfur-containing amino acids; this gives rise to the odor of rotten eggs and contributes to the odor of flatulence.

As organic matter decays under low-oxygen (or hypoxic) conditions (such as in swamps, eutrophic lakes or dead zones of oceans), sulfate-reducing bacteria will use the sulfates present in the water to oxidize the organic matter, producing hydrogen sulfide as waste. Some of the hydrogen sulfide will react with metal ions in the water to produce metal sulfides, which are not water soluble. These metal sulfides, such as ferrous sulfide FeS, are often black or brown, leading to the dark colour of sludge.

Several groups of bacteria can use hydrogen sulfide as fuel, oxidizing it to elemental sulfur or to sulfate by using dissolved oxygen, metal oxides (e.g., Fe oxyhydroxides and Mn oxides) or nitrate as oxidant. The purple sulfur bacteria and the green sulfur bacteria use hydrogen sulfide as electron donor in photosynthesis, thereby producing elemental sulfur. (In fact, this mode of photosynthesis is older than the mode of cyanobacteria, algae, and plants, which uses water as electron donor and liberates oxygen.)

Mass Extinctions

Hydrogen sulfide has been implicated in some of the several mass extinctions that have occurred in the Earth's past. In particular, a buildup of hydrogen sulfide in the atmosphere may have caused the Permian-Triassic extinction event 252 million years ago.

Organic residues from these extinction boundaries indicate that the oceans were anoxic (oxygen-depleted) and had species of shallow plankton that metabolized H_2S. The formation of H_2S may have been initiated by massive volcanic eruptions, which emitted carbon dioxide and methane into the atmosphere, which warmed the oceans, lowering their capacity to absorb oxygen that would otherwise oxidize H_2S. The increased levels of hydrogen sulfide could have killed oxygen-generating plants as well as depleted the ozone layer, causing further stress. Small H_2S blooms have been detected in modern times in the Dead Sea and in the Atlantic ocean off the coast of Namibia.

Cooling Tower

Cooling towers are heat removal devices used to transfer process waste heat to the atmosphere. Cooling towers may either use the evaporation of water to remove process heat and cool the working fluid to near the wet-bulb air temperature or in the case of *closed circuit dry cooling towers* rely solely on air to cool the working fluid to near the dry-bulb air temperature.

Common applications include cooling the circulating water used in oil refineries, chemical plants, power stations and building cooling. The towers vary in size from small roof-top units to very large hyperboloid structures that can be up to 200 metres tall and 100 metres in diameter, or rectangular structures that can be over 40 metres tall and 80 metres long.

Smaller towers are normally factory-built, while larger ones are constructed on site. They are often associated with nuclear power plants in popular culture, although cooling towers are constructed on many types of buildings.

A hyperboloid cooling tower was patented by Frederik van Iterson and Gerard Kuypers in 1918. The first hyperboloid cooling towers were built in the last 1920s in Liverpool, England to cool water used at a electrical power station that used coal.

Classification by Use

HVAC

An HVAC (heating, ventilating, and air conditioning) cooling tower is a subcategory rejecting heat from a chiller. Water-cooled chillers are normally more energy efficient than air-cooled chillers due to heat rejection to tower water at or near wet-bulb temperatures. Air-cooled chillers must reject heat at the dry-bulb temperature, and thus have a lower average reverse-Carnot cycle effectiveness. Large office buildings, hospitals, and schools typically use one or more cooling towers as part of their air conditioning systems. Generally, industrial cooling towers are much larger than HVAC towers.

HVAC use of a cooling tower pairs the cooling tower with a water-cooled chiller or water-cooled condenser. A *ton* of air-conditioning is the removal of 12,000 Btu/hour (3500 W). The *equivalent ton* on the cooling tower side actually rejects about 15,000 Btu/hour (4400 W) due to the heat-equivalent of the energy needed to drive the chiller's compressor. This *equivalent ton* is defined as the heat rejection in cooling 3 U.S. gallons/minute (1,500 pound/hour) of water 10 °F (6 °C), which amounts to 15,000 Btu/hour, or a chiller coefficient of performance (COP) of 4.0. This COP is equivalent to an energy efficiency ratio (EER) of 14.

Cooling towers are also used in HVAC systems that have multiple water source heat pumps that share a common piping *water loop*. In this type of system, the water circulating inside the water loop removes heat from the condenser of the heat pumps whenever the heat pumps are working in the cooling mode, then the cooling tower is used to remove heat from the water loop and reject it to the atmosphere. When the heat pumps are working in heating mode, the condensers draw heat out of the loop water and reject it into the space to be heated.

Industrial Cooling Towers

Industrial cooling towers can be used to remove heat from various sources such as machinery or heated process material. The primary use of large, industrial cooling towers is to remove the heat absorbed in the circulating cooling water systems used in power plants, petroleum refineries, petrochemical plants, natural gas processing plants, food processing plants, semi-conductor plants, and for other industrial facilities such as in condensers of distillation columns, for cooling liquid in crystallization, etc.

The circulation rate of cooling water in a typical 700 MW coal-fired power plant with a cooling tower amounts to about 71,600 cubic metres an hour (315,000 U.S. gallons per minute) and the circulating water requires a supply water make-up rate of perhaps 5 percent (i.e., 3,600 cubic metres an hour).

If that same plant had no cooling tower and used once-through cooling water, it would require about 100,000 cubic metres an hour and that amount of water would have to be continuously returned to the ocean, lake or river from which it was obtained and continuously re-supplied to the plant.

Furthermore, discharging large amounts of hot water may raise the temperature of the receiving river or lake to an unacceptable level for the local ecosystem. Elevated water temperatures can kill fish and other aquatic organisms. A cooling tower serves to dissipate the heat into the atmosphere instead and wind and air diffusion spreads the heat over a much larger area than hot water can distribute heat in a body of water. Some coal-fired and nuclear power plants located in coastal areas do make use of once-through ocean water. But even there, the offshore discharge water outlet requires very careful design to avoid environmental problems.

Petroleum refineries also have very large cooling tower systems. A typical large refinery processing 40,000 metric tonnes of crude oil per day (300,000 barrels (48,000 m^3) per day) circulates about 80,000 cubic metres of water per hour through its cooling tower system.

The world's tallest cooling tower is the 200 metres tall cooling tower of Niederaussem Power Station.

Classification by Build

Package Type

This type of cooling towers are preassembled and can be simply transported on trucks as they are compact machines. The capacity of package type towers is limited and for that reason, they are usually preferred by facilities with low heat rejection requirements such as food processing plants, textile plants, buildings like hospitals, hotels, malls, chemical processing plants, automotive factories etc.

Due to intensive use in domestic areas, sound level control is a relatively more important issue for package type cooling towers.

Field Erected Type

Such facilities as power plants, steel processing plants, petroleum refineries, petrochemical plants prefer field erected type cooling towers due to their large requirements of heat rejection. These towers are a lot larger in size compared to the package type cooling towers. A typical field erected cooling tower has pultruded FRP structure, FRP cladding, a mechanical unit for air draft, drift eliminator and fill.

Heat Transfer Methods

With respect to the heat transfer mechanism employed, the main types are:

- *Wet cooling towers* or simply *open circuit cooling towers* operate on the principle of evaporation. The working fluid and the evaporated fluid (usually water) are one and the same.
- *Dry cooling towers* operate by heat transfer through a surface that separates the working fluid from ambient air, such as in a tube to air heat exchanger, utilizing convective heat transfer. They do not use evaporation.
- *Fluid coolers* or *closed circuit cooling towers* are hybrids that pass the working fluid through a tube bundle, upon which clean water is sprayed and a fan-induced draft applied. The resulting heat transfer performance is much closer to that of a wet cooling tower, with the advantage provided by a dry cooler of protecting the working fluid from environmental exposure and contamination.

In a wet cooling tower (or open circuit cooling tower), the warm water can be cooled to a temperature lower than the ambient air dry-bulb temperature, if the air is relatively dry. As ambient air is drawn past a flow of water, a small portion of the water evaporate, the energy required by that portion of the water to evaporate is taken from the remaining mass of water reducing his temperature (approximately by 970 BTU for each pound of evaporated water). Evaporation results in saturated air conditions, lowering the temperature of the water process by the tower to a value close to wet bulb air temperature, which is lower than the ambient dry bulb air temperature, the difference determined by the humidity of the ambient air.

To achieve better performance (more cooling), a medium called *fill* is used to increase the surface area and the time of contact between the air and water flows.

Splash fill consists of material placed to interrupt the water flow causing splashing. *Film fill* is composed of thin sheets of material (usually PVC) upon which the water flows. Both methods create increased surface area and time of contact between the fluid (water) and the gas (air).

Air Flow Generation Methods

With respect to drawing air through the tower, there are three types of cooling towers:

- *Natural draft*, which utilizes buoyancy via a tall chimney. Warm, moist air *naturally* rises due to the density differential to the dry, cooler outside air. Warm moist air is less dense than drier air at the same pressure. This moist air buoyancy produces a current of air through the tower.
- *Mechanical draft*, which uses power driven fan motors to force or draw air through the tower.
  - *Induced draft*: A mechanical draft tower with a fan at the discharge which pulls air through tower. The fan *induces* hot moist air out the discharge. This produces low entering and high exiting air velocities, reducing the possibility of *recirculation* in which discharged air flows back into the air intake. This fan/fin arrangement is also known as *draw-through*.
  - *Forced draft*: A mechanical draft tower with a blower type fan at the intake. The fan *forces* air into the tower, creating high entering and low exiting air velocities. The low exiting velocity is much more susceptible to recirculation. With the fan on the air intake, the fan is more susceptible to complications due to freezing conditions. Another disadvantage is that a forced draft design typically requires more motor horsepower than an equivalent induced draft design. The forced draft benefit is its ability to work with

high static pressure. They can be installed in more confined spaces and even in some indoor situations. This fan/fill geometry is also known as *blow-through*.

- Fan assisted natural draft. A hybrid type that appears like a natural draft though airflow is assisted by a fan.

Hyperboloid (sometimes incorrectly known as hyperbolic) cooling towers have become the design standard for all natural-draft cooling towers because of their structural strength and minimum usage of material. The hyperboloid shape also aids in accelerating the upward convective air flow, improving cooling efficiency. They are popularly associated with nuclear power plants. However, this association is misleading, as the same kind of cooling towers are often used at large coal-fired power plants as well. Similarly, not all nuclear power plants have cooling towers, instead cooling their heat exchangers with lake, river or ocean water.

Categorization by Air-to-water Flow

Crossflow

Crossflow is a design in which the air flow is directed perpendicular to the water flow.

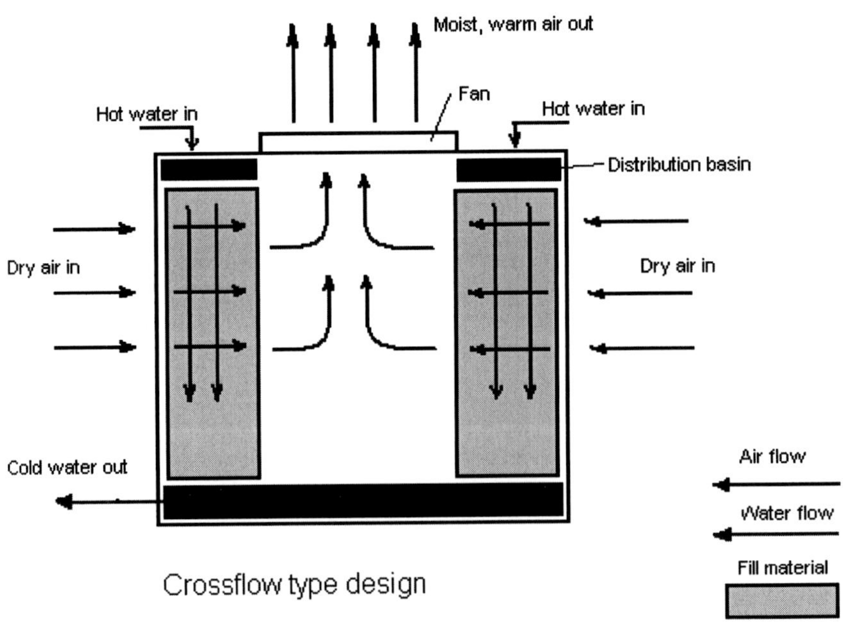

Crossflow type design

Air flow enters one or more vertical faces of the cooling tower to meet the fill material. Water flows (perpendicular to the air) through the fill by gravity.

The air continues through the fill and thus past the water flow into an open plenum area. A *distribution* or *hot water basin* consisting of a deep pan with holes or *nozzles* in the bottom is utilized in a crossflow tower. Gravity distributes the water through the nozzles uniformly across the fill material.

Counterflow

In a counterflow design the air flow is directly opposite to the water flow. Air flow first enters an open area beneath the fill media and is then drawn up vertically. The water is sprayed through pressurized nozzles and flows downward through the fill, opposite to the air flow.

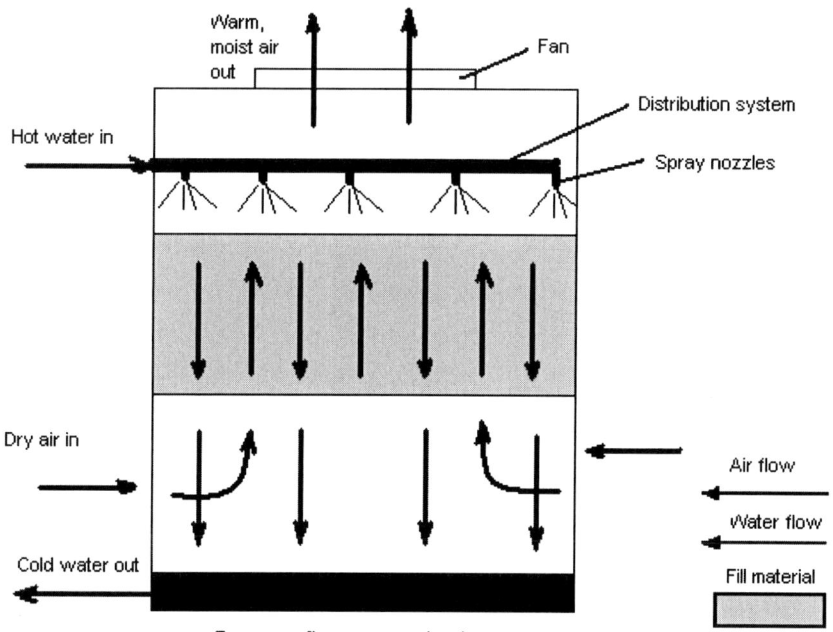

Counterflow type design

Common to both designs:

- The interaction of the air and water flow allow a partial equalization and evaporation of water.
- The air, now saturated with water vapour, is discharged from the cooling tower.

- A *collection* or *cold water basin* is used to contain the water after its interaction with the air flow.

Both crossflow and counterflow designs can be used in natural draft and mechanical draft cooling towers.

Cooling Tower as a Flue Gas Stack

At some modern power stations, equipped with flue gas purification like the Power Station Staudinger Grosskrotzenburg and the Power Station Rostock, the cooling tower is also used as a flue gas stack (industrial chimney). At plants without flue gas purification, problems with corrosion may occur.

Wet Cooling Tower Material Balance

Quantitatively, the material balance around a wet, evaporative cooling tower system is governed by the operational variables of makeup flow rate, evaporation and windage losses, draw-off rate, and the concentration cycles:

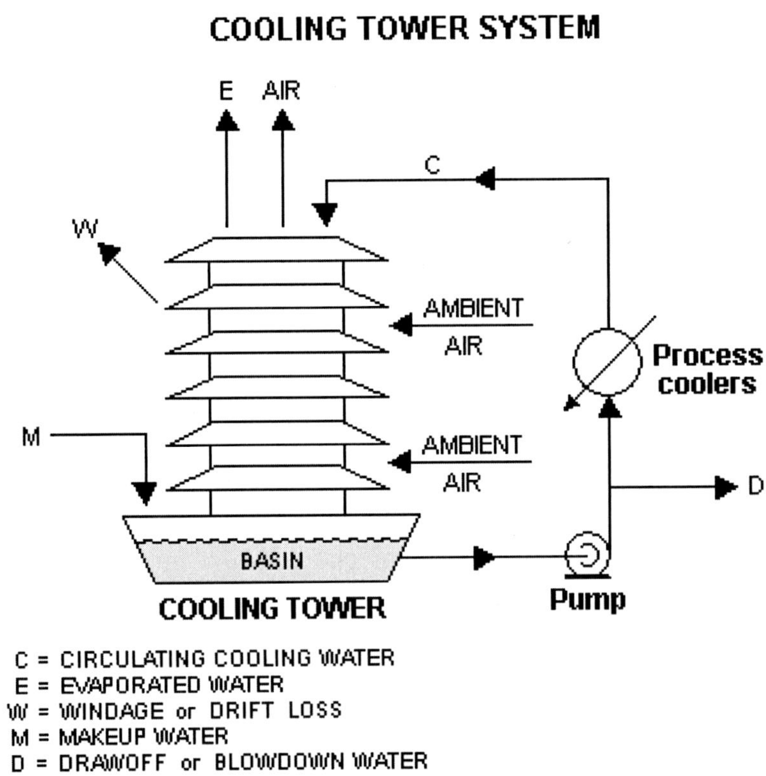

M = Make-up water in m^3/h

C = Circulating water in m^3/h

D = Draw-off water in m^3/h

E = Evaporated water in m^3/h

W = Windage loss of water in m^3/h

X = Concentration in ppmw (of any completely soluble salts ... usually chlorides)

X_M = Concentration of chlorides in make-up water (M), in ppmw

X_C = Concentration of chlorides in circulating water (C), in ppmw

Cycles = Cycles of concentration = X_C/X_M (dimensionless)

ppmw = parts per million by weight

In the above sketch, water pumped from the tower basin is the cooling water routed through the process coolers and condensers in an industrial facility.

The cool water absorbs heat from the hot process streams which need to be cooled or condensed, and the absorbed heat warms the circulating water (C).

The warm water returns to the top of the cooling tower and trickles downward over the fill material inside the tower. As it trickles down, it contacts ambient air rising up through the tower either by natural draft or by forced draft using large fans in the tower. That contact causes a small amount of the water to be lost as windage (W) and some of the water (E) to evaporate.

The heat required to evaporate the water is derived from the water itself, which cools the water back to the original basin water temperature and the water is then ready to recirculate. The evaporated water leaves its dissolved salts behind in the bulk of the water which has not been evaporated, thus raising the salt concentration in the circulating cooling water. To prevent the salt concentration of the water from becoming too high, a portion of the water is drawn off (D) for disposal. Fresh water makeup (M) is supplied to the tower basin to compensate for the loss of evaporated water, the windage loss water and the draw-off water.

A water balance around the entire system is:

$$M = E + D + W$$

Since the evaporated water (E) has no salts, a chloride balance around the system is:

$M\ (X_M) = D\ (X_C) + W\ (X_C) = X_C\ (D + W)$

and, therefore:

X_C/X_M = Cycles of concentration $= M \div (D + W) = M \div (M - E) = 1 + [E \div (D + W)]$

From a simplified heat balance around the cooling tower:

$E = C \cdot \Delta T \cdot c_p \div H_V$

where:

H_V = latent heat of vaporization of water = ca. 2260 kJ/kg

ΔT = water temperature difference from tower top to tower bottom, in °C

c_p = specific heat of water = ca. 4.184 kJ/(kg.°C)

Windage (or drift) losses (W) from large-scale industrial cooling towers, in the absence of manufacturer's data, may be assumed to be:

W = 0.3 to 1.0 percent of C for a natural draft cooling tower without windage drift eliminators

W = 0.1 to 0.3 percent of C for an induced draft cooling tower without windage drift eliminators

W = about 0.005 percent of C (or less) if the cooling tower has windage drift eliminators

Cycles of concentration represents the accumulation of dissolved minerals in the recirculating cooling water. Draw-off (or blowdown) is used principally to control the buildup of these minerals.

The chemistry of the makeup water including the amount of dissolved minerals can vary widely. Makeup waters low in dissolved minerals such as those from surface water supplies (lakes, rivers etc.) tend to be aggressive to metals (corrosive). Makeup waters from ground water supplies (wells) are usually higher in minerals and tend to be scaling (deposit minerals). Increasing the amount of minerals present in the water by cycling can make water less aggressive to piping however excessive levels of minerals can cause scaling problems.

As the cycles of concentration increase the water may not be able to hold the minerals in solution. When the solubility of these minerals have been exceeded they can precipitate out as mineral solids and cause fouling and heat exchange problems in the cooling tower or the heat exchangers.

The temperatures of the recirculating water, piping and heat exchange surfaces determine if and where minerals will precipitate from the recirculating water. Often a professional water treatment consultant will evaluate the makeup water and the operating conditions of the cooling tower and recommend an appropriate range for the cycles of concentration. The use of water treatment chemicals, pretreatment such as water softening, pH adjustment, and other techniques can affect the acceptable range of cycles of concentration.

Concentration cycles in the majority of cooling towers usually range from 3 to 7. In the United States the majority of water supplies are well waters and have significant levels of dissolved solids. On the other hand one of the largest water supplies, New York City, has a surface supply quite low in minerals and cooling towers in that city are often allowed to concentrate to 7 or more cycles of concentration.

Besides treating the circulating cooling water in large industrial cooling tower systems to minimize scaling and fouling, the water should be filtered and also be dosed with biocides and algaecides to prevent growths that could interfere with the continuous flow of the water. For closed loop evaporative towers, corrosion inhibitors may be used, but caution should be taken to meet local environmental regulations as some inhibitors use chromates.

Ambient conditions dictate the efficiency of any given tower due to the amount of water vapour the air is able to absorb and hold, as can be determined on a psychrometric chart.

Cooling Towers and Legionnaires' Disease

Another very important reason for using biocides in cooling towers is to prevent the growth of *Legionella*, including species that cause legionellosis or *Legionnaires' disease*, most notably *L. pneumophila*, or Mycobacterium avium. The various *Legionella* species are the cause of *Legionnaires' disease* in humans and transmission is via exposure to aerosols—the inhalation of mist droplets containing the bacteria. Common sources of *Legionella* include cooling towers used in open recirculating evaporative cooling water systems, domestic hot water systems, fountains, and similar disseminators that tap into a public water supply. Natural sources include freshwater ponds and creeks.

French researchers found that *Legionella* bacteria travelled up to 6 kilometres through the air from a large contaminated cooling tower at a petrochemical plant in Pas-de-Calais, France. That outbreak killed 21 of the 86 people who had a laboratory-confirmed infection.

Drift (or windage) is the term for water droplets of the process flow allowed to escape in the cooling tower discharge. Drift eliminators are used in order to hold drift rates typically to 0.001%-0.005% of the circulating flow rate. A typical drift eliminator provides multiple directional changes of airflow while preventing the escape of water droplets. A well-designed and well-fitted drift eliminator can greatly reduce water loss and potential for Legionella or other chemical exposure.

Many governmental agencies, cooling tower manufacturers and industrial trade organizations have developed design and maintenance guidelines for preventing or controlling the growth of *Legionella* in cooling towers. Below is a list of sources for such guidelines:

- Centers for Disease Control and PreventionPDF (1.35 MB) - Procedure for Cleaning Cooling Towers and Related Equipment
- Cooling Technology InstitutePDF (240 KB) - Best Practices for Control of Legionella, July, 2006
- Association of Water TechnologiesPDF (964 KB) - Legionella 2003
- California Energy CommissionPDF (194 KB) - Cooling Water Management Program Guidelines For Wet and Hybrid Cooling Towers at Power Plants
- SPX Cooling TechnologiesPDF (119 KB) - Cooling Towers Maintenance Procedures
- SPX Cooling TechnologiesPDF (789 KB) - ASHRAE Guideline 12-2000 - Minimizing the Risk of Legionellosis
- SPX Cooling TechnologiesPDF (83.1 KB) - Cooling Tower Inspection Tips
- Tower Tech Modular Cooling TowersPDF (109 KB) - Legionella Control
- GE Infrastructure Water & Process Technologies Betz DearbornPDF (195 KB) - Chemical Water Treatment Recommendations For Reduction of Risks Associated with Legionella in Open Recirculating Cooling Water Systems

Cooling Tower Fog

Under certain ambient conditions, plumes of water vapour (fog) can be seen rising out of the discharge from a cooling tower, and can be mistaken as smoke from a fire. If the outdoor air is at or near saturation, and the tower adds more water to the air, saturated air with liquid water droplets can be discharged—which is seen as fog. This phenomenon typically occurs on cool, humid days, but is rare in many climates.

This phenomenon can be prevented by decreasing the relative humidity of the saturated discharge air. For that purpose, in hybrid towers, saturated discharge air is mixed with heated low relative humidity air. Some air enters the tower above drift eliminator level, passing through heat exchangers. The relative humidity of the dry air is even more decreased instantly as being heated while entering the tower. The discharged mixture has a relatively lower relative humidity and the fog is invisible.

Cooling Tower Operation in Freezing Weather

Cooling towers with malfunctions can freeze during very cold weather. Typically, freezing starts at the corners of a cooling tower with a reduced or absent heat load. Increased freezing conditions can create growing volumes of ice, resulting in increased structural loads. During the winter, some sites continuously operate cooling towers with 40 °F (4 °C) water leaving the tower. Basin heaters, tower draindown, and other freeze protection methods are often employed in cold climates.

- Do not operate the tower unattended.
- Do not operate the tower without a heat load. This can include basin heaters and heat trace. Basin heaters maintain the temperature of the water in the tower pan at an acceptable level. Heat trace is a resistive element that runs along water pipes located in cold climates to prevent freezing.
- Maintain design water flow rate over the fill.
- Manipulate airflow to maintain water temperature above freezing point.

Some Commonly used Terms in the Cooling Tower Industry

- Drift - Water droplets that are carried out of the cooling tower with the exhaust air. Drift droplets have the same concentration of impurities as the water entering the tower. The drift rate is typically reduced by employing baffle-like devices, called drift eliminators, through which the air must travel after leaving the fill and spray zones of the tower. Drift can also be reduced by using warmer entering cooling tower temperatures.

- Blow-out - Water droplets blown out of the cooling tower by wind, generally at the air inlet openings. Water may also be lost, in the absence of wind, through splashing or misting. Devices such as wind screens, louvers, splash deflectors and water diverters are used to limit these losses.
- Plume - The stream of saturated exhaust air leaving the cooling tower. The plume is visible when water vapour it contains condenses in contact with cooler ambient air, like the saturated air in one's breath fogs on a cold day. Under certain conditions, a cooling tower plume may present fogging or icing hazards to its surroundings. Note that the water evaporated in the cooling process is "pure" water, in contrast to the very small percentage of drift droplets or water blown out of the air inlets.
- Blow-down - The portion of the circulating water flow that is removed in order to maintain the amount of dissolved solids and other impurities at an acceptable level. It may be noted that higher TDS (total dissolved solids) concentration in solution results in greater potential cooling tower efficiency. However the higher the TDS concentration, the greater the risk of scale, biological growth and corrosion. The amount of blow-down is primarily designated by the electrical conductivity of the circulating water. Biological growth, scaling and corrosion can be prevented by chemicals (Biocid, Sulfric Acid, corrosion inhibitor). On the other hand, the only way to decrease the electrical conductivity is increasing the amount of blow-down and subsequently increasing the amount of make up.
- Make up - The water that is needed to be added to the circulating water system in order to compensate water losses such as, evaporation, drift loss, blow-out, etc.
- Leaching - The loss of wood preservative chemicals by the washing action of the water flowing through a wood structure cooling tower.
- Noise - Sound energy emitted by a cooling tower and heard (recorded) at a given distance and direction. The sound is

generated by the impact of falling water, by the movement of air by fans, the fan blades moving in the structure, and the motors, gearboxes or drive belts.

- Approach - The approach is the difference in temperature between the cooled-water temperature and the entering-air wet bulb temperature (twb). Since the cooling towers are based on the principles of evaporative cooling, the maximum cooling tower efficiency depends on the wet bulb temperature of the air. The wet-bulb temperature is a type of temperature measurement that reflects the physical properties of a system with a mixture of a gas and a vapour, usually air and water vapour
- Range - The range is the temperature difference between the water inlet and water exit.
- Fill - Inside the tower, fills are added to increase contact surface as well as contact time between air and water. Thus they provide better heat transfer. The efficiency of the tower also depends on them. There are two types of fills that may be used:
  - Film type fill (causes water to spread into a thin film)
  - Splash type fill (breaks up water and interrupts its vertical progress)
- Full-Flow Filtration- Full-flow filtration continuously strains the entire system flow. For example, in a 100-ton system, the flow rate would be roughly 300 gal/min. A filter would be selected to accommodate the entire 300 gal/min flow rate. In this case, the filter typically is installed after the cooling tower on the discharge side of the pump. While this is the preferred method of filtration, for higher flow systems, it may be cost prohibitive.
- Side-Stream Filtration- Side-stream filtration, although popular, does not provide complete protection, but it can be effective. With side-stream filtration, a portion of the water is filtered continuously. This method works on the principle that continuous particle removal will keep the system clean. Manufacturers typically package side-stream filters on a skid, complete with a pump and controls. For high flow systems, this method is cost-effective.

Properly sizing a side-stream filtration system is critical to obtain satisfactory filter performance. There is some debate over how to properly size the side-stream system. Many engineers size the system to continuously filter the cooling tower basin water at a rate equivalent to 10% of the total circulation flow rate. For example, if the total flow of a system is 1,200 gal/min (a 400-ton system), a 120 gal/min side-stream system is specified.

- Cycle of concentration - Maximum allowed multiplier for the amount of miscellaneous substances in circulating water by the amount of those substances in make-up water.
- Treated timber - A structural material for cooling towers which was largely abandoned about 10 years ago. It is still used due to its low costs. The life of treated timber varies a lot, depending on the operating conditions of tower, such as frequency of stops, treatment of the circulating water, etc. Considering proper working conditions, estimated life of treated timber structural member is about 10 years.
- Pultruded FRP - A common structure material for cooling towers, known for its high corrosion resistance capabilities. Pultuded FRP is produced using pultrusion technology and became the most common structure material for cooling towers as it offers lower costs and requires less maintenance compared to reinforced concrete.

Fire Hazards

Cooling towers which are constructed in whole or in part of combustible materials can support propagating internal fires. The resulting damage can be sufficiently severe to require the replacement of the entire cell or tower structure. For this reason, some codes and standards recommend combustible cooling towers be provided with an automatic fire sprinkler system. Fires can propagate internally within the tower structure during maintenance when the cell is not in operation (such as for maintenance or construction), and even when the tower is in operation, especially those of the induced-draft type because of the existence of relatively dry areas within the towers.

Stability

Being very large structures, they are susceptible to wind damage, and several spectacular failures have occurred in the past. At Ferrybridge power station on 1 November 1965, the station was the site of a major structural failure, when three of the cooling towers collapsed owing to vibrations in 85 mph (137 km/h) winds. Although the structures had been built to withstand higher wind speeds, the shape of the cooling towers meant that westerly winds were funnelled into the towers themselves, creating a vortex. Three out of the original eight cooling towers were destroyed and the remaining five were severely damaged. The towers were rebuilt and all eight cooling towers were strengthened to tolerate adverse weather conditions. Building codes were changed to include improved structural support, and wind tunnel tests introduced to check tower structures and configuration.

European Market

The Cooling Tower market is composed of one third of "close wet" units and two thirds of "open wet" units, and reaches in 2010 a total of 190,24M€.

The market by country is splitted in 2010 as following:

Countries	***Sales Volume in M€***	***Share***
Benelux	18.70	9.8%
France	22.48	11.8%
Germany	41.40	21.8%
Greece	1.82	1.0%
Italy	20.48	10.8%
Poland	6.31	3.3%
Portugal	2.72	1.4%
Russia, Ukraine and CIS countries	16.59	8.7%
Scandinavia	4.64	2.4%
Spain	9.47	5.0%
Switzerland	5.33	2.8%
Turkey	8.50	4.5%
UK and Ireland	13.66	7.2%
Eastern Europe	18.16	9.5%

Steam

Steam is the technical term for water vapour, the gaseous phase of water, which is formed when water boils. In common language it is often used to refer to the visible mist of water droplets formed as this water vapour condenses in the presence of cooler air. At lower pressures, such as in the upper atmosphere or at the top of high mountains water boils at a lower temperature than the nominal 100 °C (212 °F) at standard temperature and pressure. If heated further it becomes superheated steam.

The enthalpy of vaporization is the energy required to turn water into the gaseous form when it increases in volume by 1,600 times at standard temperature and pressure; this change in volume can be converted into mechanical work by steam engines and steam turbines. Steam engines played a central role to the Industrial Revolution and modern steam turbines are used to generate electricity. If liquid water comes in contact with a very hot substance (such as lava, or molten metal) it can create a steam explosion. Steam explosions have been responsible for many foundry accidents and may also have been responsible for much of the damage to the plant in the Chernobyl accident.

Types of Steam and Conversion

Water vapour that includes water droplets is described as *wet steam*. As wet steam is heated further, the droplets evaporate, and at a high enough temperature (which depends on the pressure) all of the water evaporates and the system is in vapour-liquid equilibrium.

Superheated steam is steam at a temperature higher than its boiling point for the pressure which only occurs where all the water has evaporated or has been removed from the system.

Steam tables contain thermodynamic data for water/steam and are often used by engineers and scientists in design and operation of equipment where thermodynamic cycles involving steam are used. Additionally, thermodynamic phase diagrams for water/steam, such as a temperature-entropy diagram or a Mollier diagram shown in this article, may be useful. Steam charts are also used for analysing thermodynamic cycles.

Uses

Agricultural

In agriculture, steam is used for soil sterilization to avoid the use of harmful chemical agents and increase soil health.

Domestic

Steam's capacity to transfer heat is also used in the home: for cooking vegetables, steam cleaning of fabric and carpets, and heating buildings. In each case, water is heated in a boiler, and the steam carries the energy to a target object. "Steam showers" are actually low-temperature mist-generators, and do not actually use steam.

Electricity Generation (and Cogeneration)

About 90% of all electricity is generated using steam as the working fluid, nearly all by steam turbines.

In electric generation, steam is typically condensed at the end of its expansion cycle, and returned to the boiler for re-use. However in cogeneration, steam is piped into buildings through a district heating system to provide heat energy after its use in the electric generation cycle. The world's biggest steam generation system is the New York City steam system which pumps steam into 100,000 buildings in Manhattan from seven cogeneration plants.

Energy Storage

Figure 2 : Fireless steam locomotive

Despite the resemblance to a boiler, note the lack of a chimney and also how the cylinders are at the cab end, not the chimney end.

In other industrial applications steam is used for energy storage, which is introduced and extracted by heat transfer, usually through pipes. Steam is a capacious reservoir for thermal energy because of water's high heat of vaporization.

Fireless steam locomotives were steam locomotives that operated from a supply of steam stored on-board in a large tank resembling a conventional locomotive's boiler. This tank was filled by process steam, as is available in many sorts of large factory, such as paper mills. The locomotive's propulsion used pistons and connecting rods, as for a typical steam locomotive. These locomotives were mostly used in places where there was a risk of fire from a boiler's firebox, but were also used in factories that simply had a plentiful supply of steam to spare.

Lifting Gas

Owing to its low molecular mass, steam is an effective lifting gas, providing approximately 60% as much lift as helium and twice as much as hot air. It is not flammable, unlike hydrogen, and is cheap and abundant, unlike helium. The required heat, however, leads to condensation problems and requires an insulated envelope. These factors have limited its use thus far to mostly demonstration projects.

Mechanical Effort

A steam engine and steam turbines use the expansion of steam to drive a piston or turbine to perform mechanical work. The ability to return condensed steam as water-liquid to the boiler at high pressure with relatively little expenditure of pumping power is important. Condensation of steam to water often occurs at the low-pressure end of a steam turbine, since this maximizes the energy efficiency, but such wet-steam conditions have to be limited to avoid excessive turbine blade erosion. Engineers use an idealised thermodynamic cycle, the Rankine cycle, to model the behaviour of steam engines. Steam turbines are often used in the production of electricity.

Sterilization

An autoclave, which uses steam under pressure, is used in microbiology laboratories and similar environments for sterilization.

Steam in Piping

Steam is used in piping for utility lines. It is also used in jacketing and tracing of piping to maintain the uniform temperature in pipelines and vessels.

Wood Treatment

Steam is used in the process of wood bending, killing insects and increasing plasticity.

Concrete Treatment

Steam is used to accentuate drying especially in prefabricates.

Cleaning

Used in cleaning of fibres, sometimes prior to painting.

API Oil-water Separator

An API oil-water separator is a device designed to separate gross amounts of oil and suspended solids from the wastewater effluents of oil refineries, petrochemical plants, chemical plants, natural gas processing plants and other industrial sources. The name is derived from the fact that such separators are designed according to standards published by the American Petroleum Institute (API).

Description of the Design and Operation

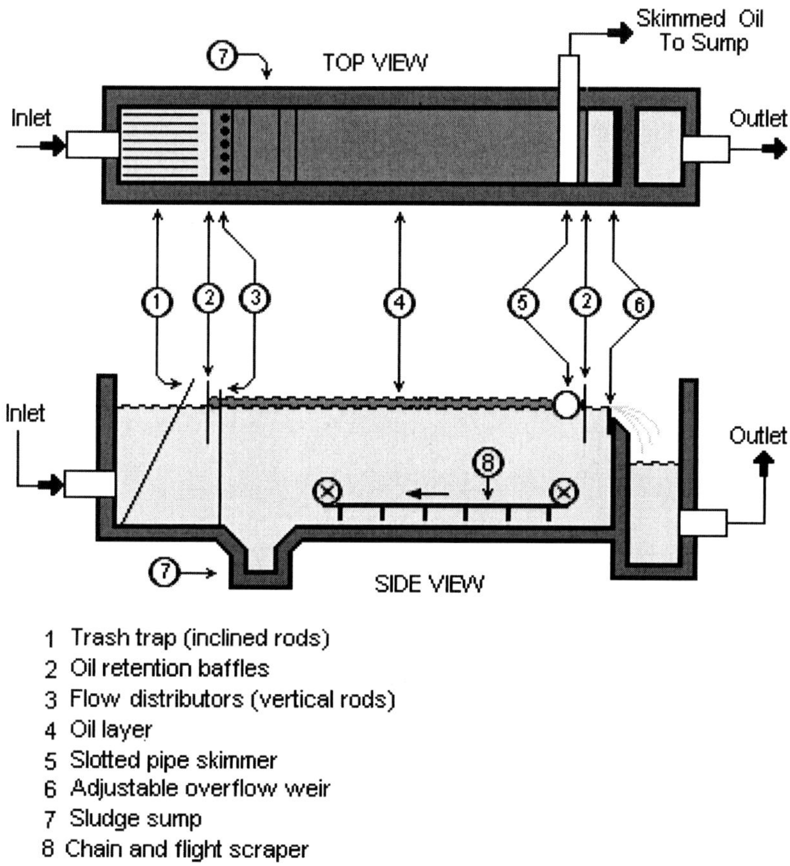

Figure 3 : A typical gravimetric API separator

The API separator is a gravity separation device designed by using Stokes Law to define the rise velocity of oil droplets based on their density and size. The design of the separator is based on the specific gravity difference between the oil and the wastewater because that difference is much smaller than the specific gravity difference between the suspended solids and water. Based on that design criterion, most of the suspended solids will settle to the bottom of the separator as a sediment layer, the oil will rise to top of the separator, and the wastewater will be the middle layer between the oil on top and the solids on the bottom.

Typically, the oil layer is skimmed off and subsequently reprocessed or disposed of, and the bottom sediment layer is removed by a chain and flight scraper (or similar device) and a sludge pump. The water layer is sent to further treatment consisting usually of a dissolved air flotation (DAF) unit for further removal of any residual oil and then to some type of biological treatment unit for removal of undesirable dissolved chemical compounds.

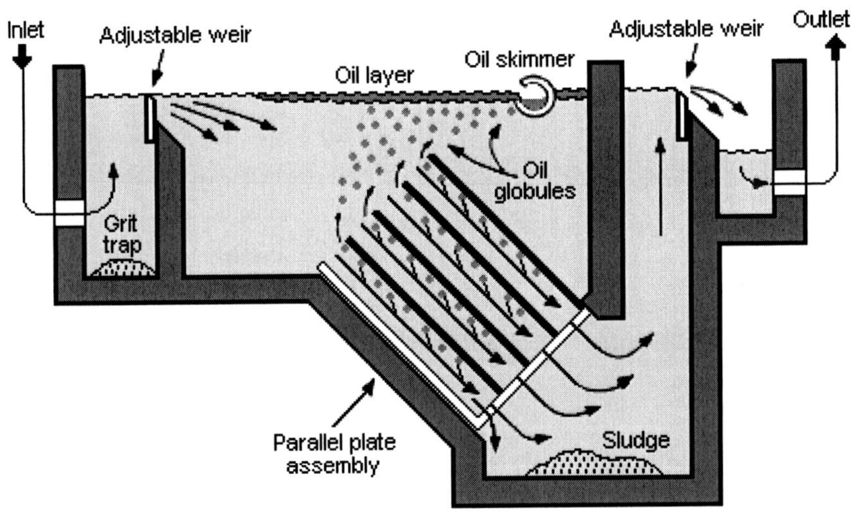

Figure 4 : A typical parallel plate separator

Parallel plate separators are similar to API separators but they include tilted parallel plate assemblies (also known as parallel packs). The underside of each parallel plate provides more surface for suspended oil droplets to coalesce into larger globules. Any sediment slides down the topside of each parallel plate. Such separators still depend upon the specific gravity between the suspended oil and the water. However, the parallel plates enhance the degree of oil-water separation. The result is that a parallel plate separator requires significantly less space than a conventional API separator to achieve the same degree of separation.

History

The API separator was developed by the API and the Rex Chain Belt Company (now USFilter Envirex Products). The first API separator was installed in 1933 at the Atlantic Refining Company (ARCO) refinery in Philadelphia. Since that time, virtually all of the refineries worldwide have installed API separators in their wastewater treatment plants. The majority of those refineries installed the API separators using the original design based on the specific gravity difference between oil and water. However, many refineries now use plastic parallel plate packing to enhance the gravity separation.

Other Oil-water Separation Applications

There are other applications requiring oil-water separation. For example:

- Oily water separators (OWS) for separating oil from the bilge water accumulated in ships as required by the international MARPOL Convention.
- Oil and water separators are commonly used in electrical substations. The transformers found in substations use a large amount of oil for cooling purposes. Moats are constructed surrounding unenclosed substations to catch any leaked oil, but these will also catch rainwater. Oil and water separators therefore provide a quicker and easier cleanup of an oil leak.

Desalter

A desalter is a process unit on an oil refinery that removes salt from the crude oil. The salt is dissolved in the water in the crude oil, not in the crude oil itself. The desalting is usually the first process in crude oil refining. The salt content after the desalter is usually measured in PTB - pounds of salt per thousand barrels of crude oil. Another specification is Basic sediment and water

The term desalter may also refer to a water desalination facility used to treat brackish water from agricultural runoff. This may be done either to produce potable water for human or animal consumption, or to reduce the salinity of river water prior to its crossing an international border, usually to comply with the terms of a treaty. Desalters are also used to treat groundwater reservoirs in areas impacted by cattle feedlots and dairies.

Desalting Crude Oil

The salts that are most frequently present in crude oil are calcium, sodium and magnesium chlorides. If these compounds are not removed from the oil several problems arise in the refining process. The high temperatures that occur downstream in the process could cause water hydrolysis, which in turn allows the formation of corrosive hydrochloric acid. Sand, silts and salt cause deposits and foul heat exchangers. The need to supply heat to vaporize water reduces crude pre-heat capacity. Sodium, arsenic and other metals can poison catalysts. By removing the suspended solids, they are not carried into the burner and eventually flue gas, where they would cause problems with environmental compliance such as flue gas opacity norms.

Grease (Lubricant)

The term grease is used to describe semisolid lubricants. Although the word *grease* is also used to describe rendered fat of animals, in the context of lubrication, grease typically applies to a material consisting of a soap emulsified with mineral or vegetable oil. The characteristic feature of greases is that they possess a high initial viscosity, which upon the application of shear, drops to give the effect of an oil-lubricated bearing of approximately the same viscosity as the base oil used in the grease.

This change in viscosity is called thixotropy. Grease is sometimes used to describe lubricating materials that are simply soft solids or high viscosity liquids, but these materials do not exhibit the shear-thinning (thixotropic) properties characteristic of the classical grease. For example, petroleum jellies such as Vaseline are not generally classified as greases. Greases are applied to mechanisms that can only be lubricated infrequently and where a lubricating oil would not stay in position. They also act as sealants to prevent ingress of water and incompressible materials. Grease-lubricated bearings have greater frictional characteristics due to their high viscosity.

Properties

A *true* grease consists of an oil and/or other fluid lubricant that is mixed with a thickener, typically a *soap*, to form a solid or semisolid.

Greases are a type of *shear-thinning* or pseudo-plastic fluid, which means that the viscosity of the fluid is reduced under shear. After sufficient force to shear the grease has been applied, the viscosity drops and approaches that of the base lubricant, such as the mineral oil. This sudden drop in shear force means that grease is considered a plastic fluid, and the reduction of shear force with time makes it thixotropic. It is often applied using a grease gun, which applies the grease to the part being lubricated under pressure, forcing the solid grease into the spaces in the part.

Thickeners

Soaps are the most common emulsifying agent used, and the selection of the type of soap is determined by the application. Soaps include calcium stearate, sodium stearate, lithium stearate, as well as mixtures of these components. Fatty acids derivatives other than stearates are also used, especially lithium 12-hydroxystearate. The nature of the soaps influences the temperature resistance (relating to the viscosity), water resistance, and chemical stability of the resulting grease.

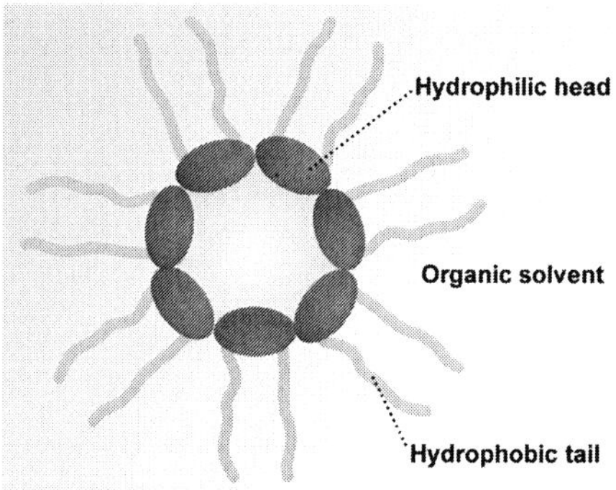

Figure 5 : An inverse micelle formed when a soap is dispersed in an oil. This structure is broken reversibly upon shearing the grease.

Powdered Solid Greases

Powdered solids may also be used as thickeners, especially as clays, which are used in some inexpensive, low performance greases. Fatty oil-based greases have also been prepared with other thickeners, such tar, graphite, or mica, which also increase the durability of the grease.

Engineering Assessment and Analysis of Greases

Lithium-based greases are the most commonly used; sodium and lithium based greases have higher melting point (dropping point) than calcium-based greases but are not resistant to the action of water. Lithium-based grease has a dropping point at 190 to 220 °C (350 to 400 °F). However the maximum usable temperature for lithium-based grease is 120 °C.

The amount of grease in a sample can be determined in a laboratory by extraction with a solvent followed by e.g. gravimetric determination.

Additives

Teflon is added to some greases to improve their lubricating properties. Gear greases consist of rosin oil, thickened with lime and mixed with mineral oil, with some percentage of water. Special-purpose greases contain glycerol and sorbitan esters. They are used, for example, in low-temperature conditions. Some greases are labeled "EP", which indicates "extreme pressure".

Under high pressure or shock loading, normal grease can be compressed to the extent that the greased parts come into physical contact, causing friction and wear. EP grease contains solid lubricants, usually graphite and/or molybdenum disulfide, to provide protection under heavy loadings.

The solid lubricants bond to the surface of the metal, and prevent metal-to-metal contact and the resulting friction and wear when the lubricant film gets too thin.

Copper is added to some greases for high pressure applications, or where corrosion could prevent dis-assembly of components later in their service life. Copaslip is the registered trademark of one such grease produced by Molyslip Atlantic Ltd, and has become a generic term (often misspelled as "copperslip" or "coppaslip") for anti-seize lubricants which contain copper.

History

Venom from the early Egyptian or Roman eras are thought to have been prepared by combining lime with olive oil. The lime saponifies some of the triglyceride that comprises oil to give a calcium grease. In the middle of the 19th century, soaps were intentionally added as thickeners to oils. Over the centuries, all manner of materials have been employed as greases. For example, black slugs *Arion ater* were used as axle-grease to lubricate wooden axle-trees or carts in Sweden.

Other Greases—Silicone Grease

Silicone grease is an amorphous fumed, silica-thickened, polysiloxane-based compound, which can be used to provide lubrication and corrosion resistance. Since it is not oil-based, it is often used where oil-based lubricants would attack rubber seals. Silicone greases also maintain stability under high temperatures. They are often used, in pure form or mixed with zinc oxide, to join heat sinks to computer CPUs.

Fluoroether-based Grease

Fluoropolymers containing C-O-C (ether) bonds for flexibility are soft, and often used as greases in demanding environments due to their inertness. Fomblin by Solvay Solexis and Krytox by duPont are prominent examples.

Laboratory Grease

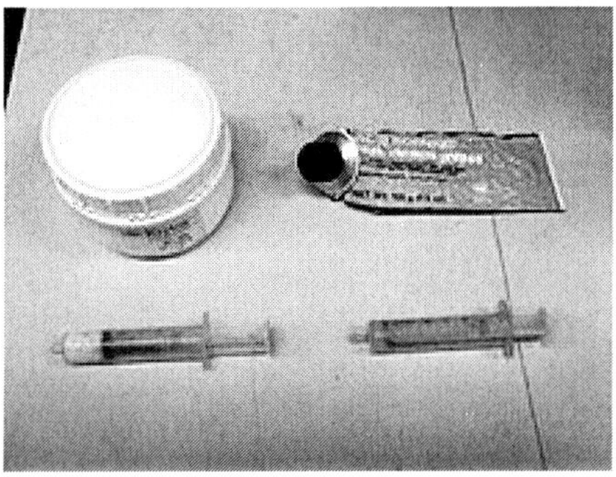

Figure 6 : Grease is used to lubricate glass stopcocks and joints. Some laboratories fill them into syringes for easy application. Two typical examples: Left - Krytox, a fluoroether-based grease; Right - a silicone-based high vacuum grease by Dow Corning.

Apiezon, silicone-based, and fluoroether-based greases are all used commonly in laboratories for lubricating stopcocks and ground glass joints.

The grease helps to prevent joints from "freezing", as well as ensuring high vacuum systems are properly sealed. Apiezon or similar hydrocarbon based greases are the cheapest, and most suitable for high vacuum applications. However, they dissolve in many organic solvents.

This quality makes clean-up with pentane or hexanes trivial, but also easily leads to contamination of reaction mixtures.

Silicone-based greases are cheaper than fluoroether-based greases. They are relatively inert and generally do not affect reactions, though reaction mixtures often get contaminated. Silicone-based greases are not easily removed with solvent, but they are removed efficiently by soaking in a base bath.

Fluoroether-based greases are inert to many substances including solvents, acids, bases, and oxidizers. They are, however, expensive, and are not easily cleaned away.

Water-soluble Grease Analogs

In some cases, the lubrication and high viscosity of a grease are desired in situations where non-toxic, non-oil based materials are required. Carboxymethyl cellulose, or CMC, is one popular material used to create a water-based analog of greases. CMC serves to both thicken the solution and add a lubricating effect, and often silicone-based lubricants are added for additional lubrication.

The most familiar example of this type of lubricant, used as a surgical and personal lubricant, is K-Y Jelly.

Viscosity

Viscosity is a measure of the resistance of a fluid which is being deformed by either shear or tensile stress. In everyday terms (and for fluids only), viscosity is "thickness" or "internal friction". Thus, water is "thin", having a lower viscosity, while honey is "thick", having a higher viscosity. Put simply, the less viscous the fluid is, the greater its ease of movement (fluidity).

Viscosity describes a fluid's internal resistance to flow and may be thought of as a measure of fluid friction. For example, high-viscosity felsic magma will create a tall, steep stratovolcano, because it cannot flow far before it cools, while low-viscosity mafic lava will create a wide, shallow-sloped shield volcano.

All real fluids (except superfluids) have some resistance to stress and therefore are viscous, but a fluid which has no resistance to shear stress is known as an ideal fluid or inviscid fluid. The study of flowing matter is known as rheology, which includes viscosity and related concepts.

Etymology

The word "viscosity" is derived from the Latin "viscum alba", meaning white mistletoe. A viscous glue called birdlime was made from mistletoe berries and was used for lime-twigs to catch birds.

Properties and Behaviour

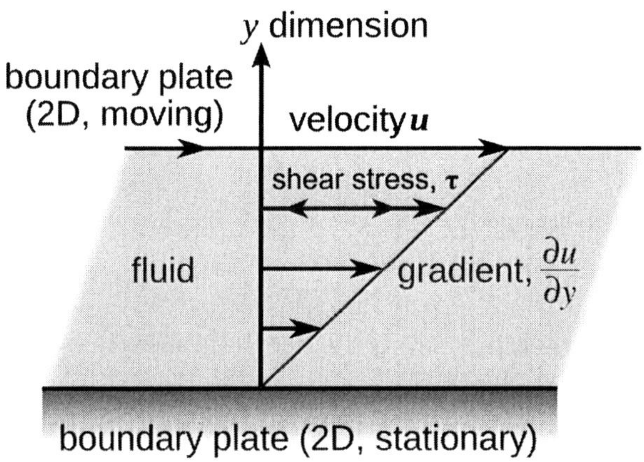

Laminar shear of fluid between two plates. Friction between the fluid and the moving boundaries causes the fluid to shear. The force required for this action is a measure of the fluid's viscosity. This type of flow is known as a Couette flow.

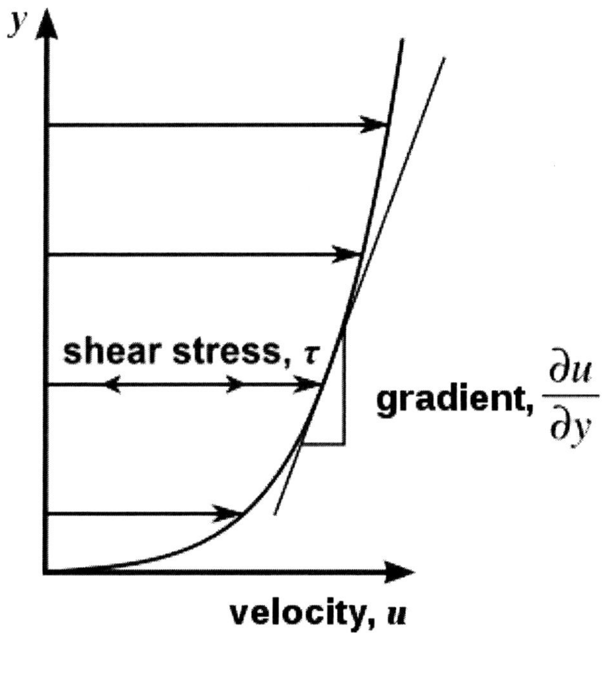

Laminar shear, the non-constant gradient, is a result of the geometry the fluid is flowing through (e.g. a pipe).

In general, in any flow, layers move at different velocities and the fluid's viscosity arises from the shear stress between the layers that ultimately opposes any applied force. The relationship between the shear stress and the velocity gradient can be obtained by considering two plates closely spaced at a distance y, and separated by a homogeneous substance. Assuming that the plates are very large, with a large area A, such that edge effects may be ignored, and that the lower plate is fixed, let a force F be applied to the upper plate. If this force causes the substance between the plates to undergo shear flow with a velocity gradient u/y (as opposed to just shearing elastically until the shear stress in the substance balances the applied force), the substance is called a fluid.

The applied force is proportional to the area and velocity gradient in the fluid and inversely proportional to the distance between the plates. Combining these three relations results in the equation:

$$F = \mu A \frac{u}{y}.$$

where μ is the proportionality factor called *dynamic viscosity*.

This equation can be expressed in terms of shear stress $\tau = \frac{F}{A}$.

Thus as expressed in differential form by Isaac Newton for straight, parallel and uniform flow, the shear stress between layers is proportional to the velocity gradient in the direction perpendicular to the layers:

$$\tau = \mu \frac{\partial u}{\partial y}.$$

Hence, through this method, the relation between the shear stress and the velocity gradient can be obtained.

Note that the *rate of shear deformation* is $\frac{u}{y}$ which can be also written as a *shear velocity*, $\frac{du}{dy}$.

James Clerk Maxwell called viscosity *fugitive elasticity* because of the analogy that elastic deformation opposes shear stress in solids, while in viscous fluids, shear stress is opposed by *rate* of deformation.

Types of Viscosity

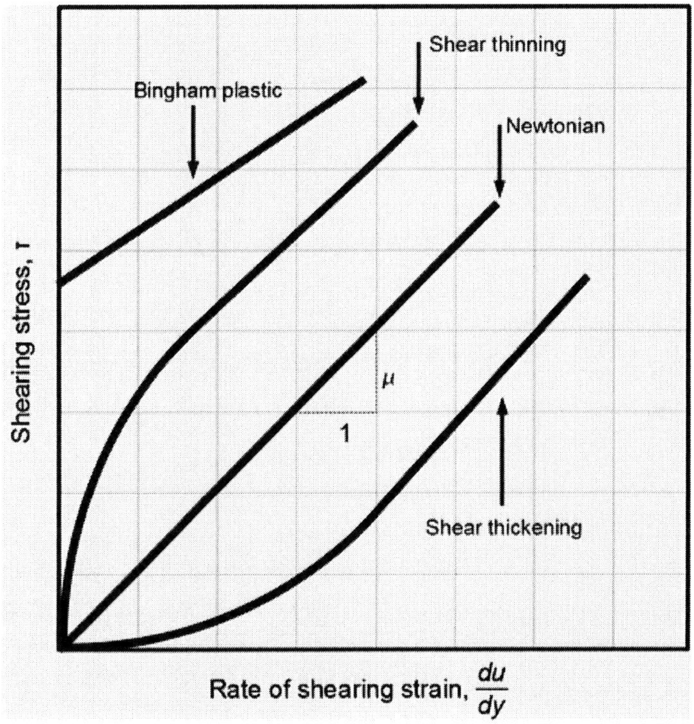

Figure 7 : Viscosity, the slope of each line, varies among materials

Newton's law of viscosity, given above, is a constitutive equation (like Hooke's law, Fick's law, Ohm's law). It is not a fundamental law of nature but an approximation that holds in some materials and fails in others.

Non-Newtonian fluids exhibit a more complicated relationship between shear stress and velocity gradient than simple linearity. Thus there exist a number of forms of viscosity:

- Newtonian: fluids, such as water and most gases which have a constant viscosity.
- Shear thickening: viscosity *increases* with the rate of shear.
- Shear thinning: viscosity *decreases* with the rate of shear. Shear thinning liquids are very commonly, but misleadingly, described as thixotropic.

- Thixotropic: materials which become *less* viscous over time when shaken, agitated, or otherwise stressed.
- Rheopectic: materials which become *more* viscous over time when shaken, agitated, or otherwise stressed.
- A Bingham plastic is a material that behaves as a solid at low stresses but flows as a viscous fluid at high stresses.
- A magnetorheological fluid is a type of "smart fluid" which, when subjected to a magnetic field, greatly increases its apparent viscosity, to the point of becoming a viscoelastic solid.

Viscosity Coefficients

Viscosity coefficients can be defined in two ways:

- Dynamic viscosity, also absolute viscosity, the more usual one (typical units Pa·s, Poise, P);
- Kinematic viscosity is the *dynamic viscosity* divided by the density (typical units cm^2/s, Stokes, St).

Viscosity is a tensorial quantity that can be decomposed in different ways into two independent components. The most usual decomposition yields the following viscosity coefficients:

- Shear viscosity, the most important one, often referred to as simply viscosity, describing the reaction to applied shear stress; simply put, it is the ratio between the pressure exerted on the surface of a fluid, in the lateral or horizontal direction, to the change in velocity of the fluid as you move down in the fluid (this is what is referred to as a velocity gradient).
- Volume viscosity (also called bulk viscosity or second viscosity) becomes important only for such effects where fluid compressibility is essential. Examples would include shock waves and sound propagation. It appears in the Stokes' law (sound attenuation) that describes propagation of sound in Newtonian liquid.

Alternatively,

- Extensional viscosity, a linear combination of shear and bulk viscosity, describes the reaction to elongation, widely used for characterizing polymers. For example, at room temperature, water has a dynamic shear viscosity of about 1.0×10^{-3} Pa·s and motor oil of about 250×10^{-3} Pa·s.

Viscosity Measurement

Viscosity is measured with various types of viscometers and rheometers. A rheometer is used for those fluids which cannot be defined by a single value of viscosity and therefore require more parameters to be set and measured than is the case for a viscometer. Close temperature control of the fluid is essential to accurate measurements, particularly in materials like lubricants, whose viscosity can double with a change of only 5 °C.

For some fluids, viscosity is a constant over a wide range of shear rates (Newtonian fluids). The fluids without a constant viscosity (non-Newtonian fluids) cannot be described by a single number. Non-Newtonian fluids exhibit a variety of different correlations between shear stress and shear rate.

One of the most common instruments for measuring kinematic viscosity is the glass capillary viscometer. In paint industries, viscosity is commonly measured with a Zahn cup, in which the efflux time is determined and given to customers. The efflux time can also be converted to kinematic viscosities (centistokes, cSt) through the conversion equations. Also used in paint, a Stormer viscometer uses load-based rotation in order to determine viscosity. The viscosity is reported in Krebs units (KU), which are unique to Stormer viscometers.

A Ford viscosity cup measures the rate of flow of a liquid. This, under ideal conditions, is proportional to the kinematic viscosity.

Vibrating viscometers can also be used to measure viscosity. These models such as the *Dynatrol* use vibration rather than rotation to measure viscosity.

Extensional viscosity can be measured with various rheometers that apply extensional stress. Volume viscosity can be measured with an acoustic rheometer. Apparent viscosity is a calculation derived from tests performed on drilling fluid used in oil or gas well development. These calculations and tests help engineers develop and maintain the properties of the drilling fluid to the specifications required.

Units

Dynamic Viscosity

The usual symbol for dynamic viscosity used by mechanical and chemical engineers — as well as fluid dynamicists — is the Greek letter mu (μ). The symbol η is also used by chemists, physicists, and the IUPAC.

The SI physical unit of dynamic viscosity is the pascal-second (Pa·s), (equivalent to N·s/m^2, or kg/(m·s)).

If a fluid with a viscosity of one Pa·s is placed between two plates, and one plate is pushed sideways with a shear stress of one pascal, it moves a distance equal to the thickness of the layer between the plates in one second. Water at 20 °C has a viscosity of 0.001002 Pa·s.

The cgs physical unit for dynamic viscosity is the *poise* (P), named after Jean Louis Marie Poiseuille. It is more commonly expressed, particularly in ASTM standards, as *centipoise* (cP). Water at 20 °C has a viscosity of 1.0020 cP.

1 P = 0.1 Pa·s,

1 cP = 1 mPa·s = 0.001 Pa·s.

Kinematic Viscosity

In many situations, we are concerned with the ratio of the inertial force to the viscous force (i.e. the Reynolds number, $Re = VD/\nu$), the former characterized by the fluid density ρ. This ratio is characterized by the *kinematic viscosity* (Greek letter nu, ν), defined as follows:

$$\nu = \frac{\mu}{\rho}$$

The SI unit of ν is m^2/s. The SI unit of ρ is kg/m^3.

The cgs physical unit for kinematic viscosity is the *stokes* (St), named after George Gabriel Stokes. It is sometimes expressed in terms of *centiStokes* (cSt). In U.S. usage, *stoke* is sometimes used as the singular form.

1 St = 1 cm^2·s^{-1} = 10^{-4} m^2·s^{-1}.

1 cSt = 1 mm^2·s^{-1} = 10^{-6}m^2·s^{-1}.

Water at 20 °C has a kinematic viscosity of about 1 cSt.

The kinematic viscosity is sometimes referred to as diffusivity of momentum, because it is analogous to diffusivity of heat and diffusivity of mass. It is therefore used in dimensionless numbers which compare the ratio of the diffusivities.

Fluidity

The reciprocal of viscosity is *fluidity*, usually symbolized by $\varphi = 1 / \mu$ or $F = 1 / \mu$, depending on the convention used, measured in *reciprocal poise* (cm·s·g^{-1}), sometimes called the *rhe*. *Fluidity* is seldom used in engineering practice.

The concept of fluidity can be used to determine the viscosity of an ideal solution. For two components *a* and *b*, the fluidity when *a* and *b* are mixed is

$$F \approx \chi_a F_a + \chi_b F_b,$$

which is only slightly simpler than the equivalent equation in terms of viscosity:

$$\mu \approx \frac{1}{\chi_a / \mu_a + \chi_b / \mu_b},$$

where χ_a and χ_b is the mole fraction of component *a* and *b* respectively, and μ_a and μ_b are the components pure viscosities.

Non-standard Units

The Reyn is a British unit of dynamic viscosity.

Viscosity index is a measure for the change of kinematic viscosity with temperature. It is used to characterise lubricating oil in the automotive industry.

At one time the petroleum industry relied on measuring kinematic viscosity by means of the Saybolt viscometer, and expressing kinematic viscosity in units of *Saybolt Universal Seconds* (SUS). Other abbreviations such as SSU (*Saybolt Seconds Universal*) or SUV (*Saybolt Universal Viscosity*) are sometimes used. Kinematic viscosity in centistoke can be converted from SUS according to the arithmetic and the reference table provided in ASTM D 2161.

Molecular Origins

The viscosity of a system is determined by how molecules constituting the system interact. There are no simple but correct expressions for the viscosity of a fluid. The simplest exact expressions are the Green–Kubo relations for the linear shear viscosity or the Transient Time Correlation Function expressions derived by Evans and Morriss in 1985. Although these expressions are each exact in order to calculate the viscosity of a dense fluid, using these relations requires the use of molecular dynamics computer simulations.

Gases

Viscosity in gases arises principally from the molecular diffusion that transports momentum between layers of flow. The kinetic theory of gases allows accurate prediction of the behaviour of gaseous viscosity.

Within the regime where the theory is applicable:

- Viscosity is independent of pressure and
- Viscosity increases as temperature increases.

James Clerk Maxwell published a famous paper in 1866 using the kinetic theory of gases to study gaseous viscosity. To understand why the viscosity is independent of pressure, consider two adjacent boundary layers (A and B) moving with respect to each other.

The internal friction (the viscosity) of the gas is determined by the probability a particle of layer A enters layer B with a corresponding transfer of momentum.

Maxwell's calculations showed him that the viscosity coefficient is proportional to both the density, the mean free path and the mean velocity of the atoms. On the other hand, the *mean free path* is inversely proportional to the density. So an increase of pressure doesn't result in any change of the viscosity.

Relation to Mean Free Path of Diffusing Particles

In relation to diffusion, the kinematic viscosity provides a better understanding of the behavior of mass transport of a dilute species. Viscosity is related to shear stress and the rate of shear in a fluid, which illustrates its dependence on the mean free path, λ, of the diffusing particles.

From fluid mechanics, for a Newtonian fluid, the shear stress, τ, on a unit area moving parallel to itself, is found to be proportional to the rate of change of velocity with distance perpendicular to the unit area:

$$\tau = \mu \frac{du_x}{dy}$$

for a unit area parallel to the x-z plane, moving along the x axis. We will derive this formula and show how μ is related to λ.

Interpreting shear stress as the time rate of change of momentum, p, per unit area A (rate of momentum flux) of an arbitrary control surface gives;

$$\tau = \frac{\dot{p}}{A} = \frac{\dot{m}\langle u_x \rangle}{A}.$$

where $\langle u_x \rangle$ is the average velocity along x of fluid molecules hitting the unit area, with respect to the unit area.

$$\dot{m} = \rho \bar{u} A$$

$\langle u_x \rangle = \frac{1}{2}\lambda \frac{du_x}{dy}$, assuming that molecules hitting the unit area come from all distances between 0 and λ (equally distributed), and that their average velocities change linearly with distance (always true for small enough λ). From this follows:

$$\tau = \underbrace{\frac{1}{2}\rho \bar{u} \lambda \cdot \frac{du_x}{dy} \Rightarrow \nu = \frac{\mu}{\rho}}_{\mu} = \tfrac{1}{2}\bar{u}\lambda,$$

where

$\dot{m}$ is the rate of fluid mass hitting the surface,

ρ is the density of the fluid,

$\bar{u}$ is the average molecular speed ($\bar{u} = \sqrt{\langle u^2 \rangle}$),

μ is the dynamic viscosity.

Effect of Temperature on the Viscosity of a Gas

Sutherland's formula can be used to derive the dynamic viscosity of an ideal gas as a function of the temperature:

$$\mu = \mu_0 \frac{T_0 + C}{T + C}\left(\frac{T}{T_0}\right)^{3/2}.$$

This in turn is equal to

$$\lambda \frac{T^{3/2}}{T + C}, \quad \text{where} \quad \lambda = \frac{\mu_0 (T_0 + C)}{T_0^{3/2}} \quad \text{is a constant.}$$

in Sutherland's formula:

- μ = dynamic viscosity in (Pa·s) at input temperature T,
- μ_0 = reference viscosity in (Pa·s) at reference temperature T_0,
- T = input temperature in kelvins,
- T_0 = reference temperature in kelvins,
- C = Sutherland's constant for the gaseous material in question.

Valid for temperatures between $0 < T < 555$ K with an error due to pressure less than 10% below 3.45 MPa.

Viscosity of a Dilute Gas

The Chapman-Enskog equation may be used to estimate viscosity for a dilute gas. This equation is based on a semi-theoretical assumption by Chapman and Enskog. The equation requires three empirically determined parameters: the collision diameter (σ), the maximum energy of attraction divided by the Boltzmann constant (ε/κ) and the collision integral ($\omega(T^*)$).

$$\mu_0 \times 10^6 = 2.6693 \frac{(MT)^{1/2}}{\sigma^2 \omega(T^*)},$$

with

- $T^* = \kappa T/\varepsilon$ — reduced temperature (dimensionless),
- μ_0 = viscosity for dilute gas ($\kappa\mu$Pa.s),
- M = molecular mass (g/mol),
- T = temperature (K),
- σ = the collision diameter (Å),
- ε/κ = the maximum energy of attraction divided by the Boltzmann constant (K),
- ω_μ = the collision integral.

Liquids

In liquids, the additional forces between molecules become important. This leads to an additional contribution to the shear stress though the exact mechanics of this are still controversial. Thus, in liquids:

- Viscosity is independent of pressure (except at very high pressure); and

- Viscosity tends to fall as temperature increases (for example, water viscosity goes from 1.79 cP to 0.28 cP in the temperature range from 0 °C to 100 °C).

The dynamic viscosities of liquids are typically several orders of magnitude higher than dynamic viscosities of gases.

Viscosity of Blends of Liquids

The viscosity of the blend of two or more liquids can be estimated using the Refutas equation. The calculation is carried out in three steps. The first step is to calculate the Viscosity Blending Number (VBN) (also called the Viscosity Blending Index) of each component of the blend:

$$(1)\quad \mathrm{VBN} = 14.534 \times \ln\left[\ln(\upsilon + 0.8)\right] + 10.975$$

where υ is the kinematic viscosity in centistokes (cSt). It is important that the kinematic viscosity of each component of the blend be obtained at the same temperature.

The next step is to calculate the VBN of the blend, using this equation:

$$(2)\quad \mathrm{VBN}_{\mathrm{Blend}} = \left[x_A \times \mathrm{VBN}_A\right] + \left[x_B \times \mathrm{VBN}_B\right] + \cdots + \left[x_N \times \mathrm{VBN}_N\right]$$

where x_X is the mass fraction of each component of the blend.

Once the viscosity blending number of a blend has been calculated using equation (2), the final step is to determine the kinematic viscosity of the blend by solving equation (1) for υ:

$$(3)\quad \upsilon = \exp\left(\exp\left(\frac{\mathrm{VBN}_{\mathrm{Blend}} - 10.975}{14.534}\right)\right) - 0.8,$$

where VBN_{Blend} is the viscosity blending number of the blend.

Viscosity of Selected Substances

The viscosity of air and water are by far the two most important materials for aviation aerodynamics and shipping fluid dynamics. Temperature plays the main role in determining viscosity.

Viscosity of Air

The viscosity of air depends mostly on the temperature. At 15.0 °C, the viscosity of air is 1.78×10"5 kg/(m·s), 17.8 ìPa.s or 1.78×10"5

Pa.s.. One can get the viscosity of air as a function of temperature from the Gas Viscosity Calculator

Viscosity of Water

The dynamic viscosity of water is 8.90 × 10^{-4} Pa·s or 8.90 × 10^{-3} dyn·s/cm^2 or 0.890 cP at about 25 °C. Water has a viscosity of 0.0091 poise at 25 °C, or 1 centipoise at 20 °C.

As a function of temperature T (K): (Pa·s) = $A \times 10^{B/(T-C)}$

where A=2.414 × 10^{-5} Pa·s ; B = 247.8 K ; and C = 140 K.

Viscosity of liquid water at different temperatures up to the normal boiling point is listed below.

Temperature [°C]	*Viscosity [mPa·s]*
10	1.308
20	1.002
30	0.7978
40	0.6531
50	0.5471
60	0.4668
70	0.4044
80	0.3550
90	0.3150
100	0.2822

Viscosity of Various Materials

Viscosity of Slurry

The term slurry designs mixtures of a liquid and solid particles that retain some fluidity. The viscosity of slurry can be described as relative to the viscosity of the liquid phase:

$$\mu_s = \mu_r \cdot \mu_l,$$

where μ_s and μ_l are respectively the dynamic viscosity of the slurry and liquid (Pa·s), and μ_r is the relative viscosity (dimensionless).

Depending on the size and concentration of the solid particles, several models exist that describe the relative viscosity as a function of volume fraction x of solid particles.

In the case of extremely low concentrations of fine particles, Einstein's equation may be used:

$$\mu_r = 1 + 2.5 \cdot \phi$$

In the case of higher concentrations, a modified equation was proposed by Guth and Simha which takes into account interaction between the solid particles:

$$\mu_r = 1 + 2.5 \cdot \phi + 14.1 \cdot \phi^2$$

Further modification of this equation was proposed by Thomas from the fitting of empirical data:

$$\mu_r = 1 + 2.5 \cdot \phi + 10.05 \cdot \phi^2 + A \cdot e^{B \cdot \phi},$$

where A = *0.00273* and B = *16.6.*

In the case of very high concentrations, another empirical equation was proposed by Kitano *et al.*:

$$\mu_r = (1 - \frac{\phi}{A})^{-2},$$

where A = *0.68* for smooth spherical particles.

Viscosity of Solids

On the basis that all solids such as granite flow in response to small shear stress, some researchers have contended that substances known as amorphous solids, such as glass and many polymers, may be considered to have viscosity. This has led some to the view that solids are simply "liquids" with a very high viscosity, typically greater than 10^{12} Pa·s. This position is often adopted by supporters of the widely held misconception that glass flow can be observed in old buildings. This distortion is the result of the undeveloped glass making process of earlier eras, and not due to the viscosity of glass.

However, others argue that solids are, in general, elastic for small stresses while fluids are not. Even if solids flow at higher stresses, they are characterized by their low-stress behaviour. This distinction is muddled if measurements are continued over long time periods, such as the Pitch drop experiment. Viscosity may be an appropriate characteristic for solids in a plastic regime. The situation becomes somewhat confused as the term *viscosity* is sometimes used for solid materials, for example Maxwell materials, to describe the relationship between stress and the rate of change of strain, rather than rate of shear.

These distinctions may be largely resolved by considering the constitutive equations of the material in question, which take into account both its viscous and elastic behaviours. Materials for which both their viscosity and their elasticity are important in a particular range of deformation and deformation rate are called *viscoelastic*. In geology, earth materials that exhibit viscous deformation at least three times greater than their elastic deformation are sometimes called rheids.

Viscosity of Amorphous Materials

Viscous flow in amorphous materials (e.g. in glasses and melts) is a thermally activated process:

$$\mu = A \cdot e^{Q/RT},$$

where Q is activation energy, T is temperature, R is the molar gas constant and A is approximately a constant.

The viscous flow in amorphous materials is characterized by a deviation from the Arrhenius-type behaviour: Q changes from a high value Q_H at low temperatures (in the glassy state) to a low value Q_L at high temperatures (in the liquid state). Depending on this change, amorphous materials are classified as either

- strong when: $Q_H - Q_L < Q_L$ or
- fragile when: $Q_H - Q_L$ e– Q_L.

The fragility of amorphous materials is numerically characterized by the Doremus' fragility ratio:

$$R_D = \frac{Q_H}{Q_L}$$

and strong material have $R_D < 2$ whereas fragile materials have $R_D \geq 2$.

The viscosity of amorphous materials is quite exactly described by a two-exponential equation:

$$\mu = A_1 \cdot T \cdot \left[1 + A_2 \cdot e^{B/RT}\right] \cdot \left[1 + C \cdot e^{D/RT}\right],$$

with constants A_1, A_2, B, C and D related to thermodynamic parameters of joining bonds of an amorphous material.

Not very far from the glass transition temperature, T_g, this equation can be approximated by a Vogel-Fulcher-Tammann (VFT) equation.

If the temperature is significantly lower than the glass transition temperature, $T < T_g$, then the two-exponential equation simplifies to an Arrhenius type equation:

$$\mu = A_L T \cdot e^{Q_H/RT}$$

with:

$$Q_H = H_d + H_m,$$

where H_d is the enthalpy of formation of broken bonds (termed configuron s) and H_m is the enthalpy of their motion. When the temperature is less than the glass transition temperature, $T < T_g$, the activation energy of viscosity is high because the amorphous materials are in the glassy state and most of their joining bonds are intact.

If the temperature is highly above the glass transition temperature, $T > T_g$, the two-exponential equation also simplifies to an Arrhenius type equation:

$$\mu = A_H T \cdot e^{Q_L/RT},$$

with:

$$Q_L = H_m.$$

When the temperature is higher than the glass transition temperature, $T > T_g$, the activation energy of viscosity is low because amorphous materials are melt and have most of their joining bonds broken which facilitates flow.

Eddy Viscosity

In the study of turbulence in fluids, a common practical strategy for calculation is to ignore the small-scale *vortices* (or *eddies*) in the motion and to calculate a large-scale motion with an *eddy viscosity* that characterizes the transport and dissipation of energy in the smaller-scale flow. Values of eddy viscosity used in modelling ocean circulation may be from 5×10^4 to 10^6 Pa·s depending upon the resolution of the numerical grid.

The Linear Viscous Stress Tensor

Viscous forces in a fluid are a function of the rate at which the fluid velocity is changing over distance. The velocity at any point r is specified by the velocity field $v(r)$. The velocity at a small distance dr from point r may be written as a Taylor series:

$$\mathbf{v}(\mathbf{r}+d\mathbf{r})=\mathbf{v}(\mathbf{r})+\frac{d\mathbf{v}}{d\mathbf{r}}d\mathbf{r}+\ldots,$$

where dv / dr is shorthand for the dyadic product of the del operator and the velocity:

$$\frac{d\mathbf{v}}{d\mathbf{r}}=\begin{bmatrix}\frac{\partial v_x}{\partial x} & \frac{\partial v_x}{\partial y} & \frac{\partial v_x}{\partial z}\\ \frac{\partial v_y}{\partial x} & \frac{\partial v_y}{\partial y} & \frac{\partial v_y}{\partial z}\\ \frac{\partial v_z}{\partial x} & \frac{\partial v_z}{\partial y} & \frac{\partial v_z}{\partial z}\end{bmatrix}.$$

This is just the Jacobian of the velocity field.

Viscous forces are the result of relative motion between elements of the fluid, and so are expressible as a function of the velocity field. In other words, the forces at r are a function of $v(r)$ and all derivatives of $v(r)$ at that point. In the case of linear viscosity, the viscous force will be a function of the Jacobian tensor alone. For almost all practical situations, the linear approximation is sufficient.

If we represent x, y, and z by indices 1, 2, and 3 respectively, the i,j component of the Jacobian may be written as “$_i v_j$ where “$_i$ is shorthand for “/”x_i. Note that when the first and higher derivative terms are zero, the velocity of all fluid elements is parallel, and there are no viscous forces. Any matrix may be written as the sum of an antisymmetric matrix and a symmetric matrix, and this decomposition is independent of coordinate system, and so has physical significance. The velocity field may be approximated as:

$$v_j(\mathbf{r}+d\mathbf{r})=v_j(\mathbf{r})+\frac{1}{2}\left(\partial_i v_j-\partial_j v_i\right)dr_i+\frac{1}{2}\left(\partial_i v_j+\partial_j v_i\right)dr_i,$$

where Einstein notation is now being used in which repeated indices in a product are implicitly summed. The second term from the right is the asymmetric part of the first derivative term, and it represents a rigid rotation of the fluid about r with angular velocity ω where:

$$\omega = \frac{1}{2}\nabla \times \mathbf{v} = \frac{1}{2}\begin{bmatrix} \partial_2 v_3 - \partial_3 v_2 \\ \partial_3 v_1 - \partial_1 v_3 \\ \partial_1 v_2 - \partial_2 v_1 \end{bmatrix}.$$

For such a rigid rotation, there is no change in the relative positions of the fluid elements, and so there is no viscous force associated with this term. The remaining symmetric term is responsible for the viscous forces in the fluid. Assuming the fluid is isotropic (i.e. its properties are the same in all directions), then the most general way that the symmetric term (the rate-of-strain tensor) can be broken down in a coordinate-independent (and therefore physically real) way is as the sum of a constant tensor (the rate-of-expansion tensor) and a traceless symmetric tensor (the rate-of-shear tensor):

$$\frac{1}{2}\left(\partial_i v_j + \partial_j v_i\right) = \underbrace{\frac{1}{3}\partial_k v_k \delta_{ij}}_{\text{rate-of-expansion tensor}} + \underbrace{\left(\frac{1}{2}\left(\partial_i v_j + \partial_j v_i\right) - \frac{1}{3}\partial_k v_k \delta_{ij}\right)}_{\text{rate-of-shear tensor}},$$

where δ_{ij} is the unit tensor. The most general linear relationship between the stress tensor σ and the rate-of-strain tensor is then a linear combination of these two tensors:

$$\sigma_{\text{visc};ij} = \zeta \partial_k v_k \delta_{ij} + \mu\left(\partial_i v_j + \partial_j v_i - \frac{2}{3}\partial_k v_k \delta_{ij}\right),$$

where ς is the coefficient of bulk viscosity (or "second viscosity") and μ is the coefficient of (shear) viscosity. The forces in the fluid are due to the velocities of the individual molecules. The velocity of a molecule may be thought of as the sum of the fluid velocity and the thermal velocity. The viscous stress tensor described above gives the force due to the fluid velocity only. The force on an area element in the fluid due to the thermal velocities of the molecules is just the hydrostatic pressure. This pressure term ($-p\ \delta ij$) must be added to the viscous stress tensor to obtain the total stress tensor for the fluid.

$$\sigma_{ij} = -p\delta_{ij} + \sigma_{\text{visc};ij}.$$

The infinitesimal force dFi on an infinitesimal area dAi is then given by the usual relationship:

$$dF_i = \sigma_{ij}\, dA_j.$$

Alkylation

Alkylation is the transfer of an alkyl group from one molecule to another. The alkyl group may be transferred as an alkyl carbocation, a free radical, a carbanion or a carbene (or their equivalents). Alkylating agents are widely used in chemistry because the alkyl group is probably the most common group encountered in organic molecules. Many biological target molecules or their synthetic precursors are composed of an alkyl chain with specific functional groups in a specific order. Selective alkylation, or adding parts to the chain with the desired functional groups, is used, especially if there is no commonly available biological precursor. Alkylation with only one carbon is termed methylation.

In oil refining contexts, alkylation refers to a particular alkylation of isobutane with olefins. It is a major aspect of the upgrading of petroleum. In medicine, alkylation of DNA is used in chemotherapy to damage the DNA of cancer cells. Alkylation is accomplished with the class of drugs called alkylating antineoplastic agents.

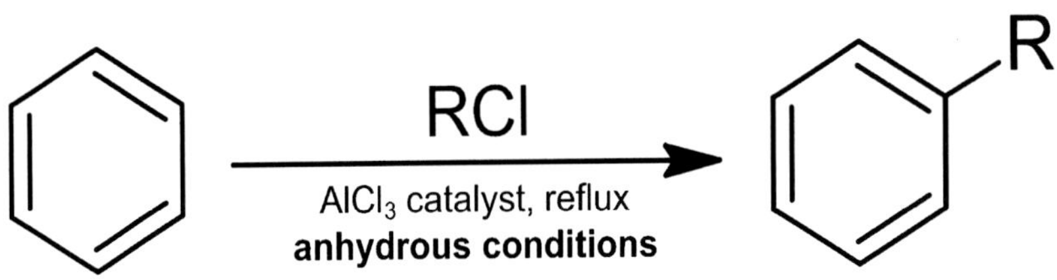

Benzene Friedel-Crafts alkylation.

Alkylating Agents

Alkylating agents are classified according to their nucleophilic or electrophilic character.

Nucleophilic Alkylating Agents

Nucleophilic alkylating agents deliver the equivalent of an alkyl anion (carbanion). Examples include the use of organometallic compounds such as Grignard (organomagnesium), organolithium, organocopper, and organosodium reagents. These compounds typically can add to an electron-deficient carbon atom such as at a carbonyl group. Nucleophilic alkylating agents can also displace halide substituents on a carbon atom. In the presence of catalysts, they also alkylate alkyl and aryl halides, as exemplified by Suzuki couplings.

Electrophilic Alkylating Agents

Electrophilic alkylating agents deliver the equivalent of an alkyl cation. Examples include the use of alkyl halides with a Lewis acid catalyst to alkylate aromatic substrates in Friedel-Crafts reactions. Alkyl halides can also react directly with amines to form C-N bonds; the same holds true for other nucleophiles such as alcohols, carboxylic acids, thiols, etc.

Electrophilic, soluble alkylating agents are often very toxic, due to their ability to alkylate DNA. They should be handled with proper PPE. This mechanism of toxicity is also responsible for the ability of some alkylating agents to perform as anti-cancer drugs in the form of alkylating antineoplastic agents, and also as chemical weapons such as mustard gas.

Alkylated DNA either does not coil or uncoil properly, or cannot be processed by information-decoding enzymes. This results in cytotoxicity with the effects of inhibition the growth of the cell, initiation of programmed cell death or apoptosis. However, mutations are also triggered, including carcinogenic mutations, explaining the higher incidence of cancer after exposure.

Alcohols and phenols can be alkylated to give alkyl ethers:

$$\text{R-OH} + \text{R'-X} \rightarrow \text{R-O-R'} + \text{H-X}$$

The produced acid HX is removed with a base, or, alternatively, the alcohol is deprotonated first to give an alkoxide or phenoxide. For example, dimethyl sulfate alkylates the sodium salt of phenol to give anisole, the methyl ether of phenol. The dimethyl sulfate is dealkylated to sodium methylsulfate.

$$\text{Ph-O}^- \text{ Na}^+ + \text{Me}_2\text{SO}_4 \rightarrow \text{Ph-O-Me} + \text{Na}^+ \text{ MeSO}_4^-$$

On the contrary, the alkylation of amines introduces the problem that the alkylation of an amine makes it *more* nucleophilic. Thus, when an electrophilic alkylating agent is introduced to a primary amine, it will preferentially alkylate all the way to a quaternary ammonium cation.

$$\text{R-NH}_2 \rightarrow \text{R-NH-R'} \rightarrow \text{R-N(R')}_2 \rightarrow \text{R-N(R')}_3^+$$

(alkylating agent omitted for clarity)

If the quaternary ammonium is not the desired product, more circuitious routes such as reductive amination are necessary.

Carbene Alkylating Agents

Carbenes are extremely reactive and are known to attack even unactivated C-H bonds. Carbenes can be generated by elimination of a diazo group. A metal can form a carbene equivalent called a transition metal carbene complex.

In Biology

Methylation is the most common type of alkylation, being associated with the transfer of a methyl group. Methylation is distinct from alkylation in that it is specifically the transfer of one carbon, whereas alkylation can refer to the transfer of long chain carbon groups. Methylation in nature is typically effected by vitamin B12-derived enzymes, where the methyl group is carried by cobalt. In methanogenesis, coenzyme M is methylated by tetrahydro-methanopterin.

Electrophilic compounds may alkylate different nucleophiles in the body. The toxicity, carcinogenity, and paradoxically, cancer cell-killing abilities of different DNA alkylating agents are an example.

Oil Refining

In a standard oil refinery process, isobutane is alkylated with low-molecular-weight alkenes (primarily a mixture of propene and butene) in the presence of a strong acid catalyst, either sulfuric acid or hydrofluoric acid. In an oil refinery it is referred to as a sulfuric acid alkylation unit (SAAU) or a hydrofluoric alkylation unit, (HFAU). Refinery workers may simply refer to it as the alky or alky unit. The catalyst protonates the alkenes (propene, butene) to produce reactive carbocations, which alkylate isobutane. The reaction is carried out at mild temperatures (0 and 30 °C) in a two-phase reaction. Because the reaction is exothermic, cooling is needed: SAAU plants require lower temperatures so the cooling medium needs to be chilled, for HFAU normal refinery cooling water will suffice. It is important to keep a high ratio of isobutane to alkene at the point of reaction to prevent side reactions which produces a lower octane product, so the plants have a high recycle of isobutane back to feed. The phases separate spontaneously, so the acid phase is vigorously mixed with the hydrocarbon phase to create sufficient contact surface.

The product is called alkylate and is composed of a mixture of high-octane, branched-chain paraffinic hydrocarbons (mostly isoheptane and isooctane). Alkylate is a premium gasoline blending stock because it has exceptional antiknock properties and is clean burning. Alkylate is also a key component of avgas. The octane number of the alkylate depends mainly upon the kind of alkenes used and upon operating conditions. For example, isooctane results from combining butylene with isobutane and has an octane rating of 100 by definition. There are other products in the alkylate, so the octane rating will vary accordingly.

Since crude oil generally contains only 10 to 40 percent of hydrocarbon constituents in the gasoline range, refineries use a fluid catalytic cracking process to convert high molecular weight hydrocarbons into smaller and more volatile compounds, which are then converted into liquid gasoline-size hydrocarbons. Alkylation processes transform low molecular-weight alkenes and iso-paraffin molecules into larger iso-paraffins with a high octane number.

Combining cracking, polymerization, and alkylation can result in a gasoline yield representing 70 percent of the starting crude oil. More advanced processes, such as cyclicization of paraffins and dehydrogenation of naphthenes forming aromatic hydrocarbons in a catalytic reformer, have also been developed to increase the octane rating of gasoline. Modern refinery operation can be shifted to produce almost any fuel type with specified performance criteria from a single crude feedstock.

In the entire range of refinery processes, alkylation is a very important process that enhances the yield of high-octane gasoline. However, not all refineries have an alkylation plant. The oil and gas journal annual survey of worldwide refining capacities for January 2007 lists many countries with no alkylation plants at their refineries.

Refineries examine whether it makes sense economically to install alkylation units. Alkylation units are complex, with substantial economy of scale. In addition to a suitable quantity of feedstock, the price spread between the value of alkylate product and alternate feedstock disposition value must be large enough to justify the installation. Alternative outlets for refinery alklylation feedstocks include sales as LPG, blending of C4 streams directly into gasoline and feedstocks for chemical plants. Local market conditions vary widely between plants. Variation in the RVP specification for gasoline between countries and between seasons dramatically impacts the amount of butane streams that can be blended directly into gasoline.

The transportation of specific types of LPG streams can be expensive so local disparities in economic conditions are often not fully mitigated by cross market movements of alkylation feedstocks.

The availability of a suitable catalyst is also an important factor in deciding whether to build an alkylation plant. If sulfuric acid is used, significant volumes are needed. Access to a suitable plant is required for the supply of fresh acid and the disposition of spent acid. If a sulfuric acid plant must be constructed specifically to support an alkylation unit, such construction will have a significant impact on both the initial requirements for capital and ongoing costs of operation. Alternatively it is possible to install a WSA Process unit to regenerate the spent acid. No drying of the gas takes place. This means that there will be no loss of acid, no acidic waste material and no heat is lost in process gas reheating.

The selective condensation in the WSA condenser ensures that the regenerated fresh acid will be 98% w/w even with the humid process gas. It is possible to combine spent acid regeneration with disposal of hydrogen sulfide by using the hydrogen sulfide as a fuel.

The second main catalyst option is hydrofluoric acid. Rates of consumption for HF acid in alkylation plants are much lower than for sulfuric acid. HF acid plants can process a wider range of feedstock mix with propylenes and butylenes. HF plants also produce alkylate with better octane rating than sulfuric plants. However, due to the hazardous nature of the material, HF acid is produced at very few locations and transportation must be managed rigorously.

Alkene

In organic chemistry, an alkene, olefin, or olefine is an unsaturated chemical compound containing at least one carbon-to-carbon double bond. The simplest acyclic alkenes, with only one double bond and no other functional groups, form an homologous series of hydrocarbons with the general formula C_nH_{2n}.

The simplest alkene is ethylene (C_2H_4), which has the International Union of Pure and Applied Chemistry (IUPAC) name *ethene*. Alkenes are also called *olefins* (an archaic synonym, widely used in the petrochemical industry). For bridged alkenes, the Bredt's rule states that a double bond cannot be placed at the bridgehead of a bridged ring system, unless the rings are large enough. Aromatic compounds are often drawn as cyclic alkenes, but their structure and properties are different and they are not considered to be alkenes.

Structure—Bonding

Like single covalent bonds, double bonds can be described in terms of overlapping atomic orbitals, except that, unlike a single bond (which consists of a single sigma bond), a carbon-carbon double bond consists of one sigma bond and one pi bond. This double bond is stronger than a single covalent bond (611 kJ/mol for C=C vs. 347 kJ/mol for C—C) and also shorter with an average bond length of 1.33 Angstroms (133 pm). Each carbon of the double bond uses its three sp^2 hybrid orbitals to form sigma bonds to three atoms. The unhybridized *2p* atomic orbitals, which lie perpendicular to the plane created by the axes of the three sp^2 hybrid orbitals, combine to form the pi bond. This bond lies outside the main C—C axis, with half of the bond on one side and half on the other.

Rotation about the carbon-carbon double bond is restricted because it involves breaking the pi bond, which requires a large amount of energy (264 kJ/mol in ethylene). As a consequence, substituted alkenes may exist as one of two isomers, called *cis* or *trans* isomers. More complex alkenes may be named using the E-Z notation, used to describe molecules having three or four different substituents (side groups). For example, of the isomers of butene, the two methyl groups of (*Z*)-but-2-ene (aka *cis*-2-butene) face the same side of the double bond, and in (*E*)-but2-ene (aka *trans*-2-butene) the methyl groups face the opposite side. These two isomers of butene are slightly different in their chemical and physical properties.

It is certainly not impossible to twist a double bond. In fact, a 90° twist requires an energy approximately equal to half the strength of a pi bond. The misalignment of the p orbitals is less than expected because pyramidalization takes place. *trans*-Cyclooctene is a stable strained alkene and the orbital misalignment is only 19° with a dihedral angle of 137° (normal 120°) and a degree of pyramidalization of 18°. This explains the dipole moment of 0.8 D for this compound (cis-isomer 0.4 D) where a value of zero is expected. The *trans* isomer of cycloheptene is only stable at low temperatures.

Shape

As predicted by the VSEPR model of electron pair repulsion, the molecular geometry of alkenes includes bond angles about each carbon in a double bond of about 120°. The angle may vary because of steric strain introduced by nonbonded interactions created by functional groups attached to the carbons of the double bond. For example, the C-C-C bond angle in propylene is 123.9°.

Physical Properties

The physical properties of alkenes are comparable with those of alkanes. The main differences between the two are that the acidity levels of alkenes are much higher than the ones in alkanes. The physical state depends on molecular mass (gases from ethene to butene - liquids from pentene onwards). The simplest alkenes, ethene, propene and butene are gases. Linear alkenes of approximately five to sixteen carbons are liquids, and higher alkenes are waxy solids.

Reactions

Alkenes are relatively stable compounds, but are more reactive than alkanes due to the presence of a carbon-carbon pi-bond. It is also attributed to the presence of pi-electrons in the molecule. The majority of the reactions of alkenes involve the rupture of this pi bond, forming new single bonds.

Alkenes serve as a feedstock for the petrochemical industry because they can participate in a wide variety of reactions.

Addition Reactions

Alkenes react in many addition reactions, which occur by opening up the double-bond. Most addition reactions to alkenes follow the mechanism of electrophilic addition. Examples of addition reactions are hydrohalogenation, halogenation, halohydrin formation, oxymercuration, hydroboration, dichlorocarbene addition, Simmons-Smith reaction, catalytic hydrogenation, epoxidation, radical polymerization and hydroxylation.

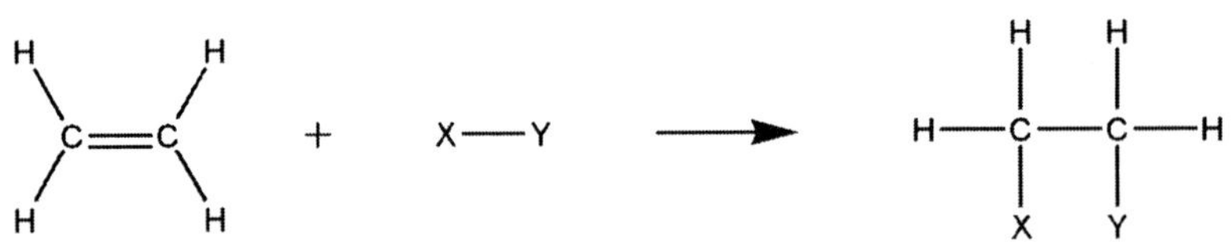

Hydrogenation

Hydrogenation of alkenes produces the corresponding alkanes. The reaction is carried out under pressure at a temperature of 200

°C in the presence of a metallic catalyst. Common industrial catalysts are based on platinum, nickel or palladium. For laboratory syntheses, Raney nickel (an alloy of nickel and aluminium) is often employed. The simplest example of this reaction is the catalytic hydrogenation of ethylene to yield ethane:

$$CH_2=CH_2 + H_2 \rightarrow CH_3\text{-}CH_3$$

Halogenation

In electrophilic halogenation the addition of elemental bromine or chlorine to alkenes yields vicinal dibromo- and dichloroalkanes (1,2-dihalides or ethylene dihalides), respectively. The decoloration of a solution of bromine in water with dichloromethylene as catalyst is an analytical test for the presence of alkenes:

$$CH_2=CH_2 + Br_2 \rightarrow BrCH_2\text{-}CH_2Br$$

It is also used as a quantitive test of unsaturation, expressed as the bromine number of a single compound or mixture. The reaction works because the high electron density at the double bond causes a temporary shift of electrons in the Br-Br bond causing a temporary induced dipole. This makes the Br closest to the double bond slightly positive and therefore an electrophile.

Hydrohalogenation

Hydrohalogenation is the addition of hydrohalic acids such as HCl or HBr to alkenes to yield the corresponding haloalkanes.

$$CH_3\text{-}CH=CH_2 + HBr \rightarrow CH_3\text{-}CHBr\text{-}CH_2\text{-}H$$

If the two carbon atoms at the double bond are linked to a different number of hydrogen atoms, the halogen is found preferentially at the carbon with fewer hydrogen substituents (Markovnikov's rule). But terminal olefin products don't yield by this method. For bromation an alternative method denominated Kharasch-Sosnovsky Reaction. is used for this purpose. It consists to add peroxides to hydrogen bromide or Copper bromide (II).

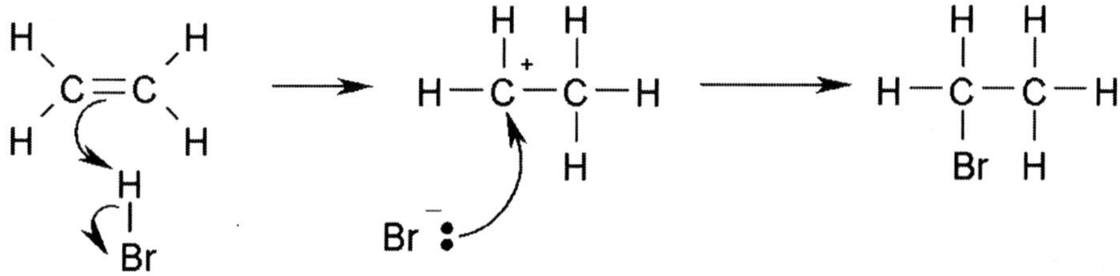

Halohydrin Formation

Alkenes react with water and halogens to form halohydrins by an addition reaction. Markovnikov regiochemistry and anti stereochemistry occur.

$$CH_2{=}CH_2 + X_2 + H_2O \rightarrow XCH_2\text{-}CH_2OH$$

Oxidation

Alkenes are oxidized with a large number of oxidizing agents. In the presence of oxygen, alkenes burn with a bright flame to produce carbon dioxide and water. Catalytic oxidation with oxygen or the reaction with percarboxylic acids yields epoxides. Reaction with ozone in ozonolysis leads to the breaking of the double bond, yielding two aldehydes or ketones. Reaction with concentrated, hot $KMnO_4$ (or other oxidizing salts) in an acidic solution will yield ketones or carboxylic acids.

$$R_1\text{-}CH{=}CH\text{-}R_2 + O_3 \rightarrow R_1\text{-}CHO + R_2\text{-}CHO + H_2O$$

This reaction can be used to determine the position of a double bond in an unknown alkene.

Oxymercuration

Hydration of alkenes via oxymercuration to produces alcohols. Reaction takes place on treatment of alkenes with strong acid as catalyst.

$$CH_2{=}CH_2 + H_2O \rightarrow CH_3\text{-}CH_2OH$$

Polymerization

Polymerization of alkenes is a reaction that yields polymers of high industrial value at great economy, such as the plastics polyethylene and polypropylene.

Polymers from alkene monomers are referred to in a general way as *polyolefins* or in rare instances as *polyalkenes*. A polymer from alpha-olefins is called a polyalphaolefin (PAO). Polymerization can proceed via either a free-radical or an ionic mechanism, converting the double to a single bond and forming single bonds to join the other monomers. Polymerization of conjugated dienes such as buta-1,3-diene or isoprene (2-methylbuta-1,3-diene) results in largely 1,4-addition with possibly some 1,2-addition of the diene monomer to a growing polymer chain.

Synthesis

Industrial Methods

Alkenes are produced by hydrocarbon cracking. Raw materials are mostly natural gas condensate components (principally ethane and propane) in the US and Mideast and naphtha in Europe and Asia. Alkanes are broken apart at high temperatures, often in the presence of a zeolite catalyst, to produce a mixture of primarily aliphatic alkenes and lower molecular weight alkanes. The mixture is feedstock dependent and separated by fractional distillation. This is mainly used for the manufacture of small alkenes (up to six carbons).

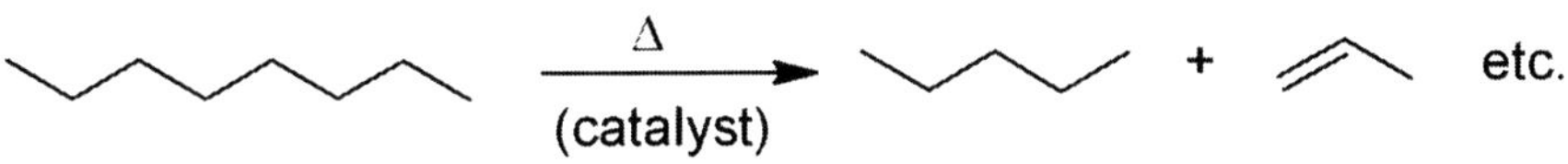

Related to this is catalytic dehydrogenation, where an alkane loses hydrogen at high temperatures to produce a corresponding alkene. This is the reverse of the catalytic hydrogenation of alkenes.

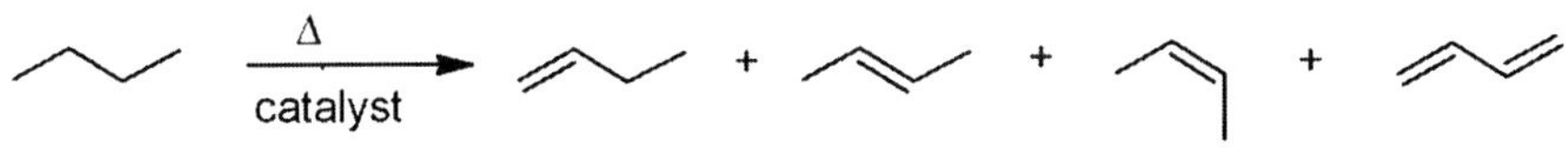

Both of these processes are endothermic, but they are driven towards the alkene at high temperatures by entropy (the TΔS portion of the equation $\Delta G = \Delta H - T\Delta S$ dominates for high T).

Catalytic synthesis of higher α-alkenes (of the type $RCH=CH_2$) can also be achieved by a reaction of ethylene with the organometallic compound triethylaluminium in the presence of nickel, cobalt, or platinum.

Elimination Reactions

One of the principal methods for alkene synthesis in the laboratory is the elimination of alkyl halides, alcohols, and similar compounds. Most common is the β-elimination via the E2 or E1 mechanism, but α-eliminations are also known.

The E2 mechanism provides a more reliable β-elimination method than E1 for most alkene syntheses. Most E2 eliminations start with an alkyl halide or alkyl sulfonate ester (such as a tosylate or triflate). When an alkyl halide is used, the reaction is called a dehydrohalogenation. For unsymmetrical products, the more substituted alkenes (those with fewer hydrogens attached to the C=C) tend to predominate. Two common methods of elimination reactions are dehydrohalogenation of alkyl halides and dehydration of alcohols. A typical example is shown below; note that the H that leaves must be *anti* to the leaving group, even though this leads to the less stable Z-isomer.

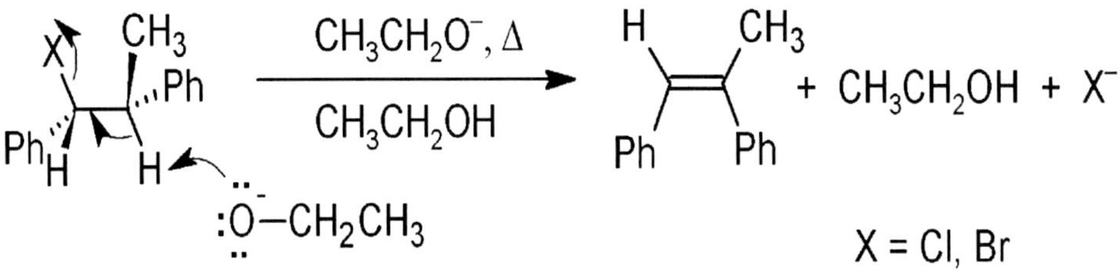

Alkenes can be synthesized from alcohols via dehydration, in which case water is lost via the E1 mechanism. For example, the dehydration of ethanol produces ethene:

$$CH_3CH_2OH + H_2SO_4 \rightarrow H_2C{=}CH_2 + H_3O^+ + HSO_4^-$$

An alcohol may also be converted to a better leaving group (e.g., xanthate), so as to allow a milder *syn*-elimination such as the Chugaev elimination and the Grieco elimination. Related reactions include eliminations by β-haloethers (the Boord olefin synthesis) and esters (ester pyrolysis).

Alkenes can be prepared indirectly from alkyl amines. The amine or ammonia is not a suitable leaving group, so the amine is first either alkylated (as in the Hofmann elimination) or oxidized to an amine oxide (the Cope reaction) to render a smooth elimination possible.

Hofmann elimination is unusual in that the *less* substituted (non-Saytseff) alkene is usually the major product. The Cope reaction is a *syn*-elimination that occurs at or below 150 °C, for example:

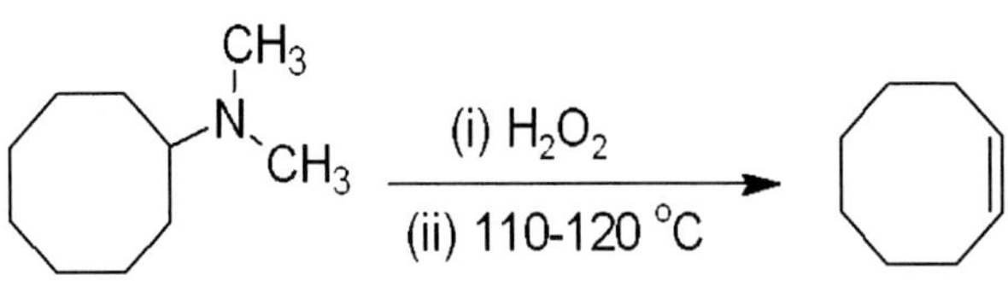

Alkenes are generated from α-halo sulfones in the Ramberg-Bäcklund reaction, via a three-membered ring sulfone intermediate.

Synthesis from Carbonyl Compounds

Another important method for alkene synthesis involves construction of a new carbon-carbon double bond by coupling of a carbonyl compound (such as an aldehyde or ketone) to a carbanion equivalent. Such reactions are sometimes called *olefinations*. The most well-known of these methods is the Wittig reaction, but other related methods are known.

The Wittig reaction involves reaction of an aldehyde or ketone with a Wittig reagent (or phosphorane) of the type $Ph_3P=CHR$ to produce an alkene and $Ph_3P=O$. The Wittig reagent is itself prepared easily from triphenylphosphine and an alkyl halide. The reaction is quite general and many functional groups are tolerated, even esters, as in this example:

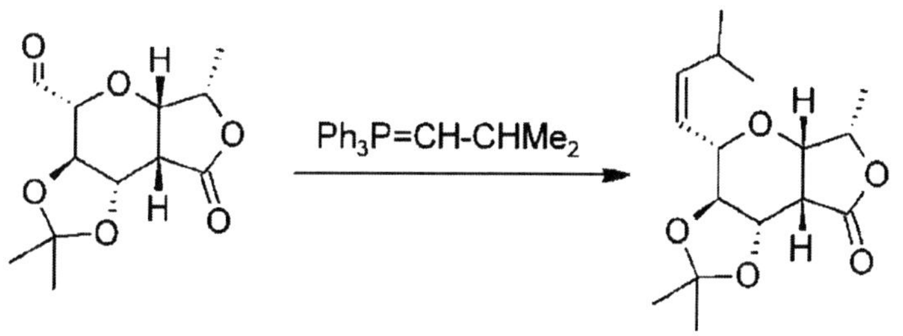

Related to the Wittig reaction is the Peterson olefination. This uses a less accessible silicon-based reagent in place of the phosphorane, but it allows for the selection of E or Z products. If an E-product is desired, another alternative is the Julia olefination, which uses the carbanion generated from a phenyl sulfone.

The Takai olefination based on an organochromium intermediate also delivers E-products. A titanium compound, Tebbe's reagent, is useful for the synthesis of methylene compounds; in this case, even esters and amides react.

A pair of carbonyl compounds can also be reductively coupled together (with reduction) to generate an alkene. Symmetrical alkenes can be prepared from a single aldehyde or ketone coupling with itself, using Ti metal reduction (the McMurry reaction). If two different ketones are to be coupled, a more complex, indirect method such as the Barton-Kellogg reaction may be used.

A single ketone can also be converted to the corresponding alkene via its tosylhydrazone, using sodium methoxide (the Bamford-Stevens reaction) or an alkyllithium (the Shapiro reaction).

Synthesis from Alkenes: Olefin Metathesis and Hydrovinylation

Alkenes can be prepared by exchange with other alkenes, in a reaction known as olefin metathesis. Frequently, loss of ethene gas is used to drive the reaction towards a desired product. In many cases, a mixture of geometric isomers is obtained, but the reaction tolerates many functional groups. The method is particularly effective for the preparation of cyclic alkenes, as in this synthesis of muscone:

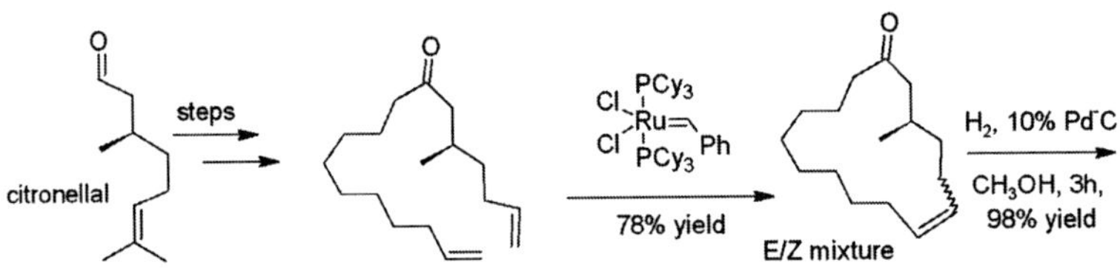

Transition metal catalyzed hydrovinylation is another important alkene synthesis process starting from alkene itself. In general, it involves the addition of a hydrogen and a vinyl group (or an alkenyl group) across a double bond.

The hydrovinylation reaction was first reported by Alderson, Jenner, and Lindsey by using rhodium and ruthenium salts, other metal catalysts commonly employed nowadays included iron, cobalt, nickel, and palladium. The addition can be done highly regio- and stereo-selectively, the choices of metal centers, ligands, substrates and counterions often play very important role.

From Alkynes

Reduction of alkynes is a useful method for the stereoselective synthesis of disubstituted alkenes. If the *cis*-alkene is desired, hydrogenation in the presence of Lindlar's catalyst (-heterogeneous catalyst that consists of palladium deposited on calcium carbonate and treated with various forms of lead) is commonly used, though hydroboration followed by hydrolysis provides an alternative approach. Reduction of the alkyne by sodium metal in liquid ammonia gives the *trans*-alkene.

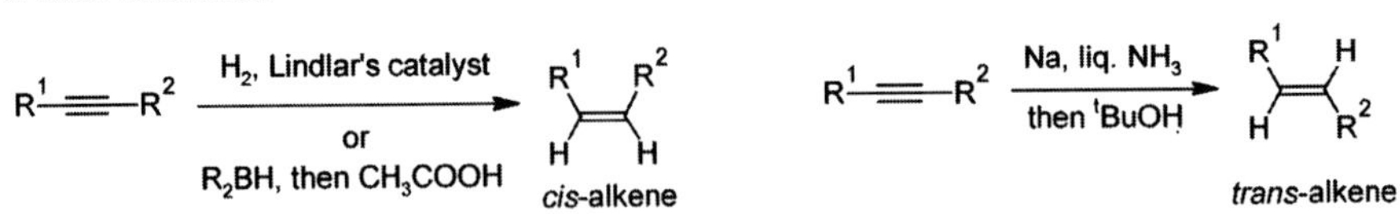

For the preparation multisubstituted alkenes, carbometalation of alkynes can give rise to a large variety of alkene derivatives.

Rearrangements and Related Reactions

Alkenes can be synthesized from other alkenes via rearrangement reactions. Besides olefin metathesis (described above), a large number of pericyclic reactions can be used such as the ene reaction and the Cope rearrangement.

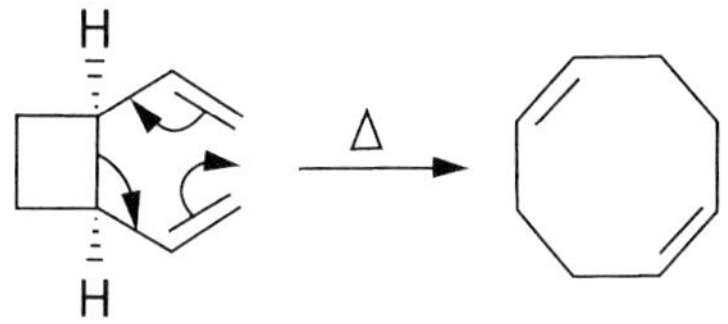

In the Diels-Alder reaction, a cyclohexene derivative is prepared from a diene and a reactive or electron-deficient alkene.

Nomenclature

IUPAC Names

To form the root of the IUPAC names for alkenes, simply change the -an- infix of the parent to -en-.

For example, CH_3-CH_3 is the alkane *ethANe*. The name of CH_2=CH_2 is therefore *ethENe*.

In higher alkenes, where isomers exist that differ in location of the double bond, the following numbering system is used:

1. Number the longest carbon chain that contains the double bond in the direction that gives the carbon atoms of the double bond the lowest possible numbers.
2. Indicate the location of the double bond by the location of its first carbon.
3. Name branched or substituted alkenes in a manner similar to alkanes.
4. Number the carbon atoms, locate and name substituent groups, locate the double bond, and name the main chain.

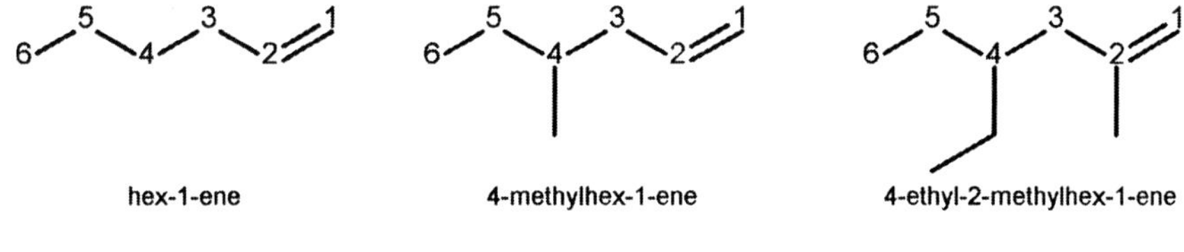

Naming substituted hex-1-enes

Cis-Trans Notation

In the specific case of disubstituted alkenes where the two carbons have one substituent each, Cis-trans notation may be used. If both substituents are on the same side of the bond, it is defined as (cis-). If the substituents are on either side of the bond, it is defined as (trans-).

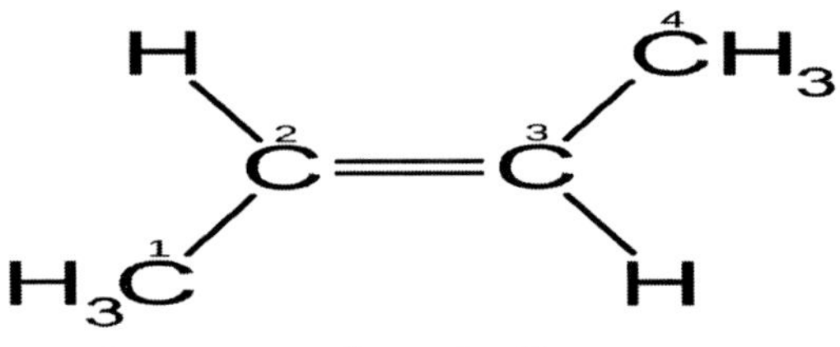

trans-but-2-ene

The difference between cis- *and* trans- *isomers.*

E,Z Notation

When an alkene has more than one substituent (especially necessary with 3 or 4 substituents), the double bond geometry is described using the labels *E* and *Z*. These labels come from the German words "entgegen," meaning "opposite," and "zusammen," meaning "together." Alkenes with the higher priority groups (as determined by CIP rules) on the same side of the double bond have these groups together and are designated *Z*. Alkenes with the higher priority groups on opposite sides are designated *E*. A mnemonic to remember this: Z notation has the higher priority groups on "ze zame zide."

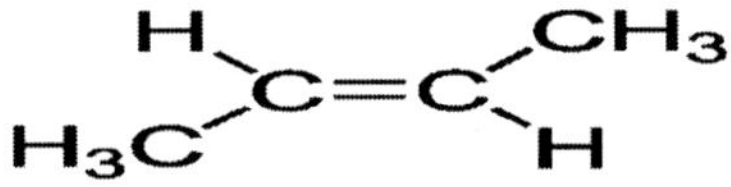

(*E*)-2-butene
entgegen (opposite)

NB. the methyl groups are on opposite sides

The difference between E *and* Z *isomers*

Groups Containing C=C Double Bonds

IUPAC recognizes two names for hydrocarbon groups containing carbon-carbon double bonds, the vinyl group and the allyl group..

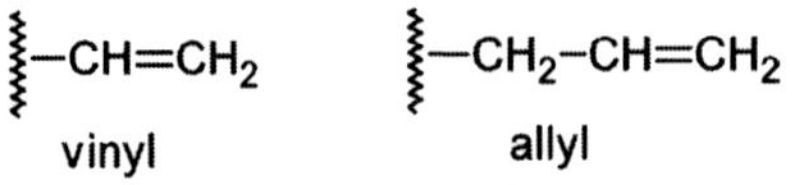

2,2,4-Trimethylpentane

2,2,4-Trimethylpentane, also known as isooctane, iso-octane, is an organic compound with the formula $(CH_3)_3CCH_2CH(CH_3)_2$. It is one of several isomers of octane (C_8H_{18}). This particular isomer is the standard for 100 point on the octane rating scale (the zero point is *n*-heptane). It is an important component of gasoline.

Production

Isooctane is produced on a massive scale in the petroleum industry by distillation of petroleum. It can also be produced from isobutylene by dimerization using an Amberlyst catalyst to produce a mixture of iso-octenes. Hydrogenation of this mixture produces 2,2,4-trimethylpentane.

Dimerization of isobutylene followed by hydrogenation produces 2,2,4-trimethylpentane.

History

Engine knocking is an unwanted process that can occur during combustion in internal combustion engines. Graham Edgar in 1926 added different amounts of *n*-heptane and 2,2,4-trimethylpentane to gasoline, and discovered that the knocking stopped when 2,2,4-trimethylpentane was added. This was the origin of the octane rating scale. The best anti-knock agent at an economical price is tetraethyllead, but because lead is very harmful to the environment, industry turned to MTBE.

Safety

In common with all hydrocarbons, inhalation or ingestion of large quantities of iso-octane is harmful. In rare cases a stronger reaction can occur. It is also flammable (as one would expect for a fuel).

2

Petrochemicals

Petrochemicals are chemical products derived from petroleum. Some chemical compounds made from petroleum are also obtained from other fossil fuels, such as coal or natural gas, or renewable sources such as corn or sugar cane. Two petrochemical classes are olefins including ethylene and propylene, and aromatics including benzene, toluene and xylene isomers. Oil refineries produce olefins and aromatics by fluid catalytic cracking of petroleum fractions. Chemical plants produce olefins by steam cracking of natural gas liquids like ethane and propane. Aromatics are produced by catalytic reforming of naphtha. Olefins and aromatics are the building blocks for a wide range of materials such as solvents, detergents, and adhesives. Olefins are the basis for polymers and oligomers used in plastics, resins, fibres, elastomers, lubricants, and gels.

Global ethylene and propylene production are ~110 million tonnes and ~65 million tonnes per annum respectively. Aromatics production is ~70 million tonnes. The largest petrochemical industries are located in the USA and Western Europe; however, major growth in new production capacity is in the Middle East and Asia. There is substantial inter-regional petrochemical trade.

Primary petrochemicals are divided into three groups depending on their chemical structure:

- Olefins includes ethylene, propylene, and butadiene. Ethylene and propylene are important sources of industrial chemicals and plastics products. Butadiene is used in making synthetic rubber.
- Aromatics includes benzene, toluene, and xylenes. Benzene is a raw material for dyes and synthetic detergents, and benzene

and toluene for isocyanates MDI and TDI used in making polyurethanes. Manufacturers use xylenes to produce plastics and synthetic fibres.

- Synthesis gas is a mixture of carbon monoxide and hydrogen used to make ammonia and methanol. Ammonia is used to make the fertilizer urea and methanol is used as a solvent and chemical intermediate.

The prefix "petro-" is an arbitrary abbreviation of the word "petroleum"; since "petro-" is Ancient Greek for "rock" and "oleum" means "oil". Therefore, the etymologically correct term would be "oleochemicals". However, the term oleochemical is used to describe chemicals derived from plant and animal fats.

Ethylene

Ethylene (IUPAC name: ethene) is a gaseous organic compound with the formula C_2H_4. It is the simplest alkene (older name: *olefin* from its oil-forming property). Because it contains a carbon-carbon double bond, ethylene is classified as an *unsaturated hydrocarbon.* Ethylene is widely used in industry and is also a plant hormone. Ethylene is the most produced organic compound in the world; global production of ethylene exceeded 107 million tonnes in 2005. To meet the ever increasing demand for ethylene, sharp increases in production facilities are added globally, particularly in the Persian Gulf countries and in China.

Structure and Bonding

This hydrocarbon has four hydrogen atoms bound to a pair of carbon atoms that are connected by a double bond. All six atoms that comprise ethylene are coplanar. The H-C-H angle is 119°, close to the 120° for ideal sp^2 hybridized carbon. The molecule is also relatively rigid: rotation about the C-C bond is a high energy process that requires breaking the π-bond.

I II

Orbital description of bonding between ethylene and a transition metal.

The π-bond in the ethylene molecule is responsible for its useful reactivity. The double bond is a region of high electron density, thus it is susceptible to attack by electrophiles. Many reactions of ethylene are catalysed by transition metals, which bind transiently to the ethylene using both the π and π* orbitals.

Uses

Major industrial reactions of ethylene include in order of scale: 1) polymerization, 2) oxidation, 3) halogenation and hydrohalogenation, 4) alkylation, 5) hydration, 6) oligomerization, and 7) hydroformylation. In the United States and Europe, approximately 90% of ethylene is used to produce three chemical compounds—ethylene oxide, ethylene dichloride, and ethylbenzene—and a variety of kinds of polyethylene.

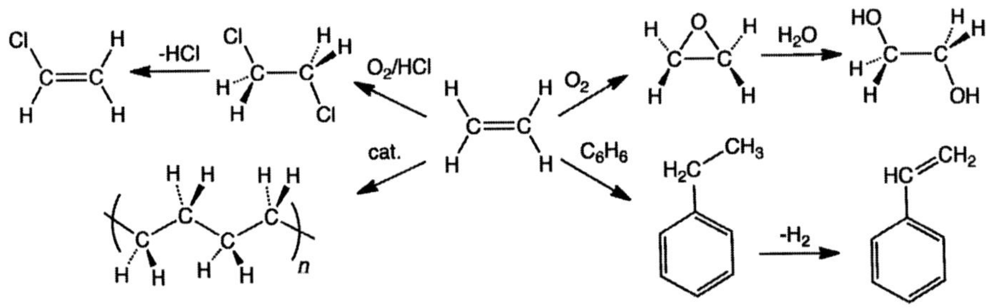

Main industrial uses of ethylene. Clockwise from the upper right: its conversions to ethylene oxide, precursor to ethylene glycol, to ethylbenzene, precursor to styrene, to various kinds of polyethylene, to ethylene dichloride, precursor to vinyl chloride.

Polymerization

Polyethylenes of various types consume more than half of world ethylene supply. Polyethylene, also called *polythene*, is the world's most widely-used plastic, being primarily used to make films used in packaging, carrier bags and trash liners. Linear alpha-olefins, produced by oligomerization (formation of short polymers) are used as precursors, detergents, plasticisers, synthetic lubricants, additives, and also as co-monomers in the production of polyethylenes.

Oxidation

Ethylene is oxidized to produce ethylene oxide, a key raw material in the production of surfactants and detergents by ethoxylation. Ethylene oxide also hydrolyzed to produce ethylene glycol, widely used as an automotive antifreeze as well as higher molecular weight glycols and glycol ethers. Ethylene undergoes oxidation by palladium to give acetaldehyde. This conversion remains a major industrial process (10M kg/y). The process proceeds via the initial complexation of ethylene to a Pd(II) centre.

Halogenation and Hydrohalogenation

Major intermediates from the halogenation and hydrohalogenation of ethylene include ethylene dichloride, ethyl chloride and ethylene dibromide. The addition of chlorine entails "oxychlorination," i.e. chlorine itself is not used. Some products derived from this group are polyvinyl chloride, trichloroethylene, perchloroethylene, methyl chloroform, polyvinylidiene chloride and copolymers, and ethyl bromide.

Alkylation

Major chemical intermediates from the alkylation with ethylene is ethylbenzene, precursor to styrene. Styrene is used principally in polystyrene for packaging and insulation, as well as in styrene-butadiene rubber for tires and footwear. On a smaller scale, ethyltoluene, ethylanilines, 1,4-hexadiene, and aluminium alkyls. Products of these intermediates include polystyrene, unsaturated polyesters and ethylene-propylene terpolymers.

Oxo-reaction

The hydroformylation (oxo-reaction) of ethylene results in propionaldehyde, a precursor to propionic acid and n-propyl alcohol.

Hydration

Ethylene can be hydrated to give ethanol, but this method is rarely used industrially.

Niche Uses

An example of a niche use is as an anesthetic agent (in an 85% ethylene/15% oxygen ratio). It can also be used to hasten fruit ripening, as well as a welding gas.

Production

In 2006, global ethylene production was 109 million tonnes. By 2010 ethylene was produced by at least 117 companies in 55 countries.

Ethylene is produced in the petrochemical industry by steam cracking. In this process, gaseous or light liquid hydrocarbons are heated to 750–950 °C, inducing numerous free radical reactions followed by immediate quench to stop these reactions. This process converts large hydrocarbons into smaller ones and introduces unsaturation. Ethylene is separated from the resulting complex mixture by repeated compression and distillation. In a related process used in oil refineries, high molecular weight hydrocarbons are cracked over zeolite catalysts. Heavier feedstocks, such as naphtha and gas oils require at least two “quench towers” downstream of the cracking furnaces to recirculate pyrolysis-derived gasoline and process water. When cracking a mixture of ethane and propane, only one water quench tower is required.

The areas of an ethylene plant are:

1. steam cracking furnaces:
2. primary and secondary heat recovery with quench;
3. a dilution steam recycle system between the furnaces and the quench system;
4. primary compression of the cracked gas (3 stages of compression);
5. hydrogen sulfide and carbon dioxide removal (acid gas removal);
6. secondary compression (1 or 2 stages);
7. drying of the cracked gas;
8. cryogenic treatment;
9. all of the cold cracked gas stream goes to the demethanizer tower. The overhead stream from the demethanizer tower consists of all the hydrogen and methane that was in the cracked gas stream. Cryogenically (“250 °F (“157 °C)) treating this overhead stream separates hydrogen from methane. Methane recovery is critical to the economical operation of an ethylene plant.
10. the bottom stream from the demethanizer tower goes to the deethanizer tower. The overhead stream from the deethanizer tower consists of all the C_2’s that were in the cracked gas stream. The C_2 stream contains acetylene, which is explosive above 200 kPa (29 psi). If the partial pressure of acetylene is expected to exceed these values, the C_2 stream is partially hydrogenated. The C_2’s then proceed to a C_2 splitter. The product ethylene is taken from the overhead of the tower and the ethane coming from the bottom of the splitter is recycled to the furnaces to be cracked again;

11. the bottom stream from the de-ethanizer tower goes to the depropanizer tower. The overhead stream from the depropanizer tower consists of all the C_3's that were in the cracked gas stream. Before feeding the C_3's to the C_3 splitter, the stream is hydrogenated to convert the methylacetylene and propadiene (allene) mix. This stream is then sent to the C_3 splitter. The overhead stream from the C_3 splitter is product propylene and the bottom stream is propane which is sent back to the furnaces for cracking or used as fuel.
12. The bottom stream from the depropanizer tower is fed to the debutanizer tower. The overhead stream from the debutanizer is all of the C_4's that were in the cracked gas stream. The bottom stream from the debutanizer (light pyrolysis gasoline) consists of everything in the cracked gas stream that is C_5 or heavier.

Since ethylene production is energy intensive, much effort has been dedicated to recovering heat from the gas leaving the furnaces. Most of the energy recovered from the cracked gas is used to make high pressure (1200 psig) steam. This steam is in turn used to drive the turbines for compressing cracked gas, the propylene refrigeration compressor, and the ethylene refrigeration compressor. An ethylene plant, once running, does not need to import steam to drive its steam turbines. A typical world scale ethylene plant (about 1.5 billion pounds of ethylene per year) uses a 45,000 horsepower (34,000 kW) cracked gas compressor, a 30,000 hp (22,000 kW) propylene compressor, and a 15,000 hp (11,000 kW) ethylene compressor.

Laboratory Aspects

Ethylene can be produced via ethanol dehydration by heating absolute ethanol with concentrated sulfuric acid or by passing ethanol vapour over hot aluminum oxide catalyst.

Interestingly for such a useful compound, ethylene is rarely used in organic synthesis in the laboratory.

Being a simple molecule, ethylene is spectroscopically simple. Its UV-vis spectrum is still used as a test of theoretical methods.

Ethylene as a Plant Hormone

Ethylene serves as a hormone in plants. It acts at trace levels throughout the life of the plant by stimulating or regulating the ripening of fruit, the opening of flowers, and the abscission (or shedding) of leaves. Commercial ripening rooms use "catalytic generators" to make ethylene gas from a liquid supply of ethanol. Typically, a gassing level of 500 ppm to 2,000 ppm is used, for 24 to 48 hours. Care must be taken to control carbon dioxide levels in ripening rooms when gassing, as high temperature ripening (68F) has been seen to produce CO2 levels of 10% in 24 hours.

History of Ethylene in Plant Biology

Ethylene has been used in practice since the ancient Egyptians, who would gash figs in order to stimulate ripening (wounding stimulates ethylene production by plant tissues). The ancient Chinese would burn incense in closed rooms to enhance the ripening of pears. In 1864, it was discovered that gas leaks from street lights led to stunting of growth, twisting of plants, and abnormal thickening of stems. In 1901, a Russian scientist named Dimitry Neljubow showed that the active component was ethylene. Doubt discovered that ethylene stimulated abscission in 1917. It wasn't until 1934 that Gane reported that plants synthesize ethylene. In 1935, Crocker proposed that ethylene was the plant hormone responsible for fruit ripening as well as senescence of vegetative tissues.

Ethylene Biosynthesis in Plants

Ethylene is produced from essentially all parts of higher plants, including leaves, stems, roots, flowers, fruits, tubers, and seedlings.

"Ethylene production is regulated by a variety of developmental and environmental factors. During the life of the plant, ethylene production is induced during certain stages of growth such as germination, ripening of fruits, abscission of leaves, and senescence of flowers. Ethylene production can also be induced by a variety of external aspects such as mechanical wounding, environmental stresses, and certain chemicals including auxin and other regulators".

The biosynthesis of the hormone starts with conversion of the amino acid methionine to S-adenosyl-L-methionine (SAM, also called Adomet) by the enzyme Met Adenosyltransferase. SAM is then converted to 1-aminocyclopropane-1-carboxylic-acid (ACC) by the enzyme ACC synthase (ACS); the activity of ACS determines the rate of ethylene production, therefore regulation of this enzyme is key for the ethylene biosynthesis. The final step requires oxygen and involves the action of the enzyme ACC-oxidase (ACO), formerly known as the

Ethylene Forming Enzyme (EFE). Ethylene biosynthesis can be induced by endogenous or exogenous ethylene. ACC synthesis increases with high levels of auxins, especially Indole acetic acid (IAA), and cytokinins. ACC synthase is inhibited by abscisic acid.

Ethylene Perception in Plants

Ethylene could be perceived by a transmembrane protein dimer complex. The gene encoding an ethylene receptor has been cloned in *Arabidopsis thaliana* and then in tomato. Ethylene receptors are encoded by multiple genes in the Arabidopsis and tomato genomes. The gene family comprises five receptors in Arabidopsis and at least six in tomato, most of which have been shown to bind ethylene. DNA sequences for ethylene receptors have also been identified in many other plant species and an ethylene binding protein has even been identified in Cyanobacteria.

Environmental and Biological Triggers of Ethylene

Environmental cues can induce the biosynthesis of the plant hormone. Flooding, drought, chilling, wounding, and pathogen attack can induce ethylene formation in the plant. In flooding, root suffers from lack of oxygen, or anoxia, which leads to the synthesis of 1-aminocyclopropane-1-carboxylic acid (ACC). ACC is transported upwards in the plant and then oxidized in leaves. The product, the ethylene causes epinasty of the leaves.

One speculation recently put forth for epinasty is the downward pointing leaves may act as pump handles in the wind. The ethylene may or may not additionally induce the growth of a valve in the xylem, but the idea would be that the plant would harness the power of the wind to pump out more water from the roots of the plants than would normally happen with transpiration.

Physiological Responses of Plants

Like the other plant hormones, ethylene is considered to have pleiotropic effects. This essentially means that it is thought that at least some of the effects of the hormone are unrelated.

What is actually caused by the gas may depend on the tissue affected as well as environmental conditions. In the evolution of plants, ethylene would simply be a message that was coopted for unrelated uses by plants during different periods of the evolutionary development.

List of Plant Responses to Ethylene

- Seedling triple response, thickening and shortening of hypocotyl with pronounced apical hook.
- In pollination, when the pollen reaches the stigma, the precursor of the ethylene, ACC, is secreted to the petal, the ACC releases ethylene with ACC oxidase.
- Stimulates leaf and flower senescence
- Stimulates senescence of mature xylem cells in preparation for plant use
- Induces leaf abscission
- Induces seed germination
- Induces root hair growth — increasing the efficiency of water and mineral absorption
- Induces the growth of adventitious roots during flooding
- Stimulates epinasty — leaf petiole grows out, leaf hangs down and curls into itself
- Stimulates fruit ripening
- Induces a climacteric rise in respiration in some fruit which causes a release of additional ethylene.
- Affects gravitropism
- Stimulates nutational bending
- Inhibits stem growth and stimulates stem and cell broadening and lateral branch growth outside of seedling stage
- Interference with auxin transport (with high auxin concentrations)
- Inhibits shoot growth and stomatal closing except in some water plants or habitually flooded ones such as some rice varieties, where the opposite occurs (conserving CO_2 and O_2)
- Induces flowering in pineapples
- Inhibits short day induced flower initiation in *Pharbitus nil* and *Chrysanthemum morifolium*.

Commercial Issues

Ethylene shortens the shelf life of many fruits by hastening fruit ripening and floral senescence. Ethylene will shorten the shelf life of cut flowers and potted plants by accelerating floral senescence and floral abscission. Flowers and plants which are subjected to stress during shipping, handling, or storage produce ethylene causing a significant reduction in floral display. Flowers affected by ethylene include carnation, geranium, petunia, rose, and many others.

Ethylene can cause significant economic losses for florists, markets, suppliers, and growers. Researchers have developed several ways to inhibit ethylene, including inhibiting ethylene synthesis and inhibiting ethylene perception. Aminoethoxyvinylglycine (AVG), Aminooxyacetic acid (AOA), and silver ions are ethylene inhibitors. Inhibiting ethylene synthesis is less effective for reducing post-harvest losses since ethylene from other sources can still have an effect. By inhibiting ethylene perception, fruits, plants and flowers don't respond to ethylene produced endogenously or from exogenous sources. Inhibitors of ethylene perception include compounds that have a similar shape to ethylene, but do not elicit the ethylene response. One example of an ethylene perception inhibitor is 1-methylcyclopropene (1-MCP).

Commercial growers of bromeliads, including pineapple plants, use ethylene to induce flowering. Plants can be induced to flower either by treatment with the gas in a chamber, or by placing a banana peel next to the plant in an enclosed area.

Chrysanthemum flowering is delayed by ethylene gas and growers have found that carbon dioxide 'burners' and the exhaust fumes from inefficient glasshouse heaters can raise the ethylene concentration to 0.05 vpm causing delay in flowering of commercial crops.

Historical Significance

Many geologists and scholars believe that the famous Greek Oracle at Delphi (the Pythia) went into her trance-like state as an effect of ethylene rising from ground faults.

History

Ethylene appears to have been discovered by Johann Joachim Becher, who obtained it by heating ethanol with sulfuric acid; he mentioned the gas in his *Physica Subterranea* (1669). Joseph Priestley also mentions the gas in his *Experiments and observations relating to the various branches of natural philosophy: with a continuation of the observations on air* (1779), where he reports that Jan Ingenhousz saw ethylene synthesized in the same way by a Mr. Enée in Amsterdam in 1777 and that Ingenhousz subsequently produced the gas himself.

The properties of ethylene were studied in 1795 by four Dutch chemists, Johann Rudolph Deimann, Adrien Paets van Troostwyck, Anthoni Lauwerenburgh and Nicolas Bondt, who found that it differed from hydrogen gas and that it contained both carbon and hydrogen. This group also discovered that ethylene could be combined with chlorine to produce the *oil of the Dutch chemists*, 1,2-dichloroethane; this discovery gave ethylene the name used for it at that time, *olefiant gas* (oil-making gas.)

In the mid-19th century, the suffix *-ene* (an Ancient Greek root added to the end of female names meaning "daughter of") was widely used to refer to a molecule or part thereof that contained one fewer hydrogen atoms than the molecule being modified. Thus, *ethylene* (C_2H_4) was the "daughter of ethyl" (C_2H_5). The name ethylene was used in this sense as early as 1852.

In 1866, the German chemist August Wilhelm von Hofmann proposed a system of hydrocarbon nomenclature in which the suffixes -ane, -ene, -ine, -one, and -une were used to denote the hydrocarbons with 0, 2, 4, 6, and 8 fewer hydrogens than their parent alkane. In this system, ethylene became *ethene*. Hofmann's system eventually became the basis for the Geneva nomenclature approved by the International Congress of Chemists in 1892, which remains at the core of the IUPAC nomenclature. However, by that time, the name ethylene was deeply entrenched, and it remains in wide use today, especially in the chemical industry.

Nomenclature

The 1979 IUPAC nomenclature rules made an exception for retaining the non-systematic name ethylene, however, this decision was reversed in the 1993 rules so the IUPAC name is now *ethene*.

Safety

Like all hydrocarbons, ethylene is an asphyxiant and combustible. It has been used as an anesthetic.

Polyethylene

Polyethylene or polythene (IUPAC name polyethene or poly(methylene)) is the most widely used plastic, with an annual production of approximately 80 million metric tons. Its primary use is within packaging (plastic bag, plastic films, geomembranes, etc.).

Description

Polyethylene is a thermoplastic polymer consisting of long chains produced by combing the ingredient monomer ethylene (IUPAC name ethene), the name comes from the ingredient and not the actual chemical resulting. The ethylene actually converts to ethane as it takes its place in a polymer and straight sections of the polymer are the same structure as the simple chain hydrocarbons, e.g., propane, decane and other straight single-bonded carbon chains.

As with any polymer, the structure of the resulting substance defies molecular description due to cross branching of the chains. The scientific name *polyethene* is systematically derived from the scientific name of the monomer.

In certain circumstances it is useful to use a structure-based nomenclature; in such cases IUPAC recommends *poly(methylene)* (poly(methanediyl) is a non-preferred alternative). The difference in names between the two systems is due to the *opening up* of the monomer's double bond upon polymerization.

The name is abbreviated to PE in a manner similar to that by which other polymers like polypropylene and polystyrene are shortened to PP and PS respectively. In the United Kingdom the polymer is commonly called polythene, although this is not recognized scientifically.

The ethene molecule (known almost universally by its common name ethylene) C_2H_4 is CH_2=CH_2, Two CH_2 groups connected by a double bond, thus:

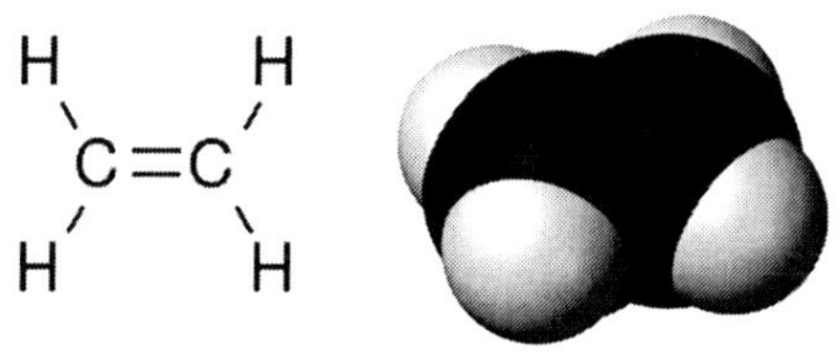

Polyethylene contains the chemical elements carbon and hydrogen.

Polyethylene is created through polymerization of ethene. It can be produced through radical polymerization, anionic addition polymerization, ion coordination polymerization or cationic addition polymerization. This is because ethene does not have any substituent groups that influence the stability of the propagation head of the polymer. Each of these methods results in a different type of polyethylene.

Classification

Polyethylene is classified into several different categories based mostly on its density and branching. The mechanical properties of PE depend significantly on variables such as the extent and type of branching, the crystal structure and the molecular weight. With regard to sold volumes, the most important polyethylene grades are HDPE, LLDPE and LDPE.

- Ultra high molecular weight polyethylene (UHMWPE)
- Ultra low molecular weight polyethylene (ULMWPE or PE-WAX)
- High molecular weight polyethylene (HMWPE)
- High density polyethylene (HDPE)
- High density cross-linked polyethylene (HDXLPE)
- Cross-linked polyethylene (PEX or XLPE)
- Medium density polyethylene (MDPE)
- Linear low density polyethylene (LLDPE)
- Low density polyethylene (LDPE)
- Very low density polyethylene (VLDPE).

UHMWPE is polyethylene with a molecular weight numbering in the millions, usually between 3.1 and 5.67 million. The high molecular weight makes it a very tough material, but results in less efficient packing of the chains into the crystal structure as evidenced by densities of less than high density polyethylene (for example, 0.930–0.935 g/cm^3). UHMWPE can be made through any catalyst technology, although Ziegler catalysts are most common. Because of its outstanding toughness and its cut, wear and excellent chemical resistance, UHMWPE is used in a diverse range of applications. These include can and bottle handling machine parts, moving parts on weaving machines, bearings, gears, artificial joints, edge protection on ice rinks and butchers' chopping boards. It competes with aramid in bulletproof vests, under the tradenames Spectra and Dyneema, and is commonly used for the construction of articular portions of implants used for hip and knee replacements.

HDPE is defined by a density of greater or equal to 0.941 g/cm^3. HDPE has a low degree of branching and thus stronger intermolecular forces and tensile strength. HDPE can be produced by chromium/silica catalysts, Ziegler-Natta catalysts or metallocene catalysts. The lack of branching is ensured by an appropriate choice of catalyst (for example, chromium catalysts or Ziegler-Natta catalysts) and reaction conditions. HDPE is used in products and packaging such as milk jugs, detergent bottles, margarine tubs, garbage containers and water pipes. One third of all toys are manufactured from HDPE. In 2007 the global HDPE consumption reached a volume of more than 30 million tons.

PEX is a medium- to high-density polyethylene containing cross-link bonds introduced into the polymer structure, changing the thermoplast into an elastomer. The high-temperature properties of the polymer are improved, its flow is reduced and its chemical resistance is enhanced. PEX is used in some potable-water plumbing systems because tubes made of the material can be expanded to fit over a metal nipple and it will slowly return to its original shape, forming a permanent, water-tight, connection.

MDPE is defined by a density range of 0.926–0.940 g/cm^3. MDPE can be produced by chromium/silica catalysts, Ziegler-Natta catalysts or metallocene catalysts. MDPE has good shock and drop resistance properties. It also is less notch sensitive than HDPE, stress cracking resistance is better than HDPE. MDPE is typically used in gas pipes and fittings, sacks, shrink film, packaging film, carrier bags and screw closures.

LLDPE is defined by a density range of 0.915–0.925 g/cm^3. LLDPE is a substantially linear polymer with significant numbers of short branches, commonly made by copolymerization of ethylene with short-chain alpha-olefins (for example, 1-butene, 1-hexene and 1-octene). LLDPE has higher tensile strength than LDPE, it exhibits higher impact and puncture resistance than LDPE. Lower thickness (gauge) films can be blown, compared with LDPE, with better environmental stress cracking resistance but is not as easy to process. LLDPE is used in packaging, particularly film for bags and sheets. Lower thickness may be used compared to LDPE.

Cable covering, toys, lids, buckets, containers and pipe. While other applications are available, LLDPE is used predominantly in film applications due to its toughness, flexibility and relative transparency. Product examples range from agricultural films, saran wrap, and bubble wrap, to multilayer and composite films. In 2009 the world LLDPE market reached a volume of almost US$24 billion (€17 billion).

LDPE is defined by a density range of 0.910–0.940 g/cm^3. LDPE has a high degree of short and long chain branching, which means that the chains do not pack into the crystal structure as well. It has, therefore, less strong intermolecular forces as the instantaneous-dipole induced-dipole attraction is less. This results in a lower tensile strength and increased ductility. LDPE is created by free radical polymerization. The high degree of branching with long chains gives molten LDPE unique and desirable flow properties. LDPE is used for both rigid containers and plastic film applications such as plastic bags and film wrap. In 2009 the global LDPE market had a volume of circa US$22.2 billion (€15.9 billion).

VLDPE is defined by a density range of 0.880–0.915 g/cm^3. VLDPE is a substantially linear polymer with high levels of short-chain branches, commonly made by copolymerization of ethylene with short-chain alpha-olefins (for example, 1-butene, 1-hexene and 1-octene). VLDPE is most commonly produced using metallocene catalysts due to the greater co-monomer incorporation exhibited by these catalysts. VLDPEs are used for hose and tubing, ice and frozen food bags, food packaging and stretch wrap as well as impact modifiers when blended with other polymers.

Recently much research activity has focused on the nature and distribution of long chain branches in polyethylene. In HDPE a relatively small number of these branches, perhaps 1 in 100 or 1,000 branches per backbone carbon, can significantly affect the rheological properties of the polymer.

Ethylene Copolymers

In addition to copolymerization with alpha-olefins, ethylene can also be copolymerized with a wide range of other monomers and ionic composition that creates ionized free radicals. Common examples include vinyl acetate (the resulting product is ethylene-vinyl acetate copolymer, or EVA, widely used in athletic-shoe sole foams) and a variety of acrylates. Applications of acrylic copolymer include packaging and sporting goods, and superplasticizer, used for cement production.

History

Polyethylene was first synthesized by the German chemist Hans von Pechmann who prepared it by accident in 1898 while heating diazomethane. When his colleagues Eugen Bamberger and Friedrich Tschirner characterized the white, waxy, substance that he had created they recognized that it contained long -CH_2- chains and termed it *polymethylene*.

The first industrially practical polyethylene synthesis was discovered (again by accident) in 1933 by Eric Fawcett and Reginald Gibson at the ICI works in Northwich, England. Upon applying extremely high pressure (several hundred atmospheres) to a mixture of ethylene and benzaldehyde they again produced a white, waxy, material. Because the reaction had been initiated by trace oxygen contamination in their apparatus, the experiment was, at first, difficult to reproduce. It was not until 1935 that another ICI chemist, Michael Perrin, developed this accident into a reproducible high-pressure synthesis for polyethylene that became the basis for industrial LDPE production beginning in 1939. Because polyethylene was found to have very low-loss properties at very high frequencies, commercial distribution in Britain was suspended on the outbreak of World War II, secrecy imposed and the new process was used to produce insulation for UHF and SHF coaxial cables of radar sets. During World War II, further research was done in the United States on the ICI process and in 1944 Bakelite Corporation at Sabine, Texas and Du Pont at Charleston, West Virginia, began large scale commercial production under license from ICI.

Subsequent landmarks in polyethylene synthesis have revolved around the development of several types of catalyst that promote ethylene polymerization at more mild temperatures and pressures. The first of these was a chromium trioxide-based catalyst discovered in 1951 by Robert Banks and J. Paul Hogan at Phillips Petroleum. In 1953 the German chemist Karl Ziegler developed a catalytic system based on titanium halides and organoaluminium compounds that worked at even milder conditions than the Phillips catalyst. The Phillips catalyst is less expensive and easier to work with, however, and both methods are used in industrial practice.

By the end of the 1950s both the Phillips- and Ziegler-type catalysts were being used for HDPE production. Phillips initially had difficulties producing a HDPE product of uniform quality and filled warehouses with off-specification plastic. However, financial ruin was unexpectedly averted in 1957 when the hula hoop, a toy consisting of a circular polyethylene tube, became a fad among youth in the United States.

A third type of catalytic system, one based on metallocenes, was discovered in 1976 in Germany by Walter Kaminsky and Hansjörg

Sinn. The Ziegler and metallocene catalyst families have since proven to be very flexible at copolymerizing ethylene with other olefins and have become the basis for the wide range of polyethylene resins available today, including very low density polyethylene and linear low-density polyethylene. Such resins, in the form of fibres like Dyneema, have (as of 2005) begun to replace aramids in many high-strength applications.

Until recently the metallocenes were the most active single-site catalysts for ethylene polymerisation known—new catalysts are typically compared to zirconocene dichloride. Much effort is currently being exerted on developing new, single-site (so-called post-metallocene) catalysts that may allow greater tuning of the polymer structure than is possible with metallocenes. Recently work by Fujita at the Mitsui corporation (amongst others) has demonstrated that certain salicylaldimine complexes of Group 4 metals show substantially higher activity than the metallocenes.

Physical Properties

Depending on the crystallinity and molecular weight, a melting point and glass transition may or may not be observable. The temperature at which these occur varies strongly with the type of polyethylene. For common commercial grades of medium- and high-density polyethylene the melting point is typically in the range 120 to 130 °C (248 to 266 °F). The melting point for average, commercial, low-density polyethylene is typically 105 to 115 °C (221 to 239 °F).

Most LDPE, MDPE and HDPE grades have excellent chemical resistance and do not dissolve at room temperature because of their crystallinity. Polyethylene (other than cross-linked polyethylene) usually can be dissolved at elevated temperatures in aromatic hydrocarbons such as toluene or xylene, or in chlorinated solvents such as trichloroethane or trichlorobenzene.

When incinerated, polyethylene burns slowly with a blue flame having a yellow tip and gives off an odour of paraffin. The material continues burning on removal of the flame source and produces a drip.

Application of Polyethylene

Polyethylene is a class of thermoplastics practically ubiquitous in consumer products. In its foam form, polyethylene is used in packaging, vibration dampening and insulation, as a barrier or buoyancy component, or as material for cushioning. Polyethylene foam is most frequently seen as a packaging material.

Polyethylene foam is buoyant, making it popular for nautical uses. Many types of polyethylene foam are approved for use in the food industry. Found in all types of packaging, polyethylene foam is used to wrap furniture, computer components, electronics, sporting goods, plants, frozen foods, clothing, bowling balls, signs, metal products, and more.

Environmental Issues

The political environment has a divided approach towards the use of plastic. As plastics are mainly based on oil or natural gas, there is a general trend, as well as political pressure, towards increased use of renewable sources. The REACH Directive regulates the handling of chemicals in Europe. The directive implies higher demands to cleansing and drives the industry towards increased use of recycled materials.

Other elements go in favour of increased use of plastics. The low weight reduces energy use and cost related to transportation compared to goods made from wood or paper. Development of advanced food packaging also prolongs shelf life of products and reduces the amount of food disposed of by the consumer. In this respect, plastics are considered to have a favourable environmental profile. However, there is no clear consensus whether or not increased use of plastics reduces CO_2 emissions.

Plastics can also be produced from renewable sources like ethylene made from, for example, sugar canes. Polyethylene is not considered biodegradable, because, except when it is exposed to UV from sunlight, it takes several centuries until it is efficiently degraded. However, in May 2008, Daniel Burd, a 16-year-old Canadian, won the Canada-Wide Science Fair in Ottawa after discovering that *Pseudomonas fluorescens*, with the help of *Sphingomonas*, two types of bacteria, can degrade over 40% of the weight of plastic bags in less than three months.

Bio-derived Polyethylene

Braskem and Toyota Tsusho Corporation started Joint marketing activities for producing polyethylene from sugar cane. Braskem will build a new facility at their existing industrial unit in Triunfo, RS,

Brazil with an annual production capacity of 200,000 short tons (180,000,000 kg), and will produce high-density polyethylene (HDPE) and low-density polyethylene (LDPE) from bioethanol derived from sugarcane.

Polyethylene can also be made from other feedstocks, including wheat grain and sugar beet.

Joining

Commonly used methods for joining include:

- Hot gas welding
- Fastening
- Infrared welding
- Laser welding
- Ultrasonic welding
- Heat sealing.

Adhesives and solvents are rarely used because polyethylene is nonpolar and has a high resistance to solvents. Pressure sensitive adhesives (PSA) are feasible if the surface is flame treated or corona treated. Commonly used adhesives include:

- Dispersion of solvent-type PSAs
- Polyurethane contact adhesives
- Two-part polyurethane or epoxy adhesives
- Vinyl acetate copolymer hot melt adhesives.

Polymerization

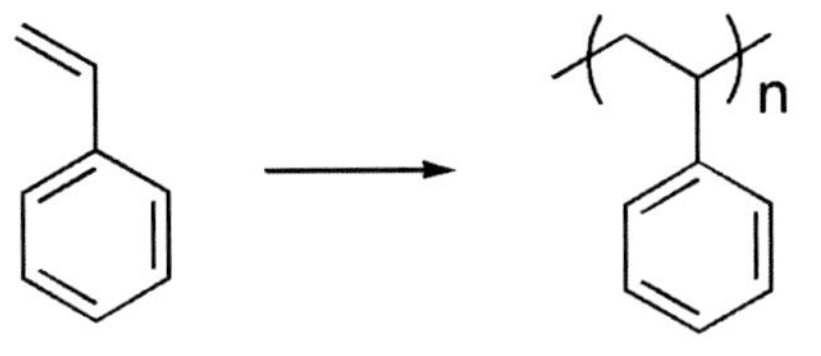

An example of alkene polymerization, in which each Styrene monomer unit's double bond reforms as a single bond with another styrene monomer and forms polystyrene.

In polymer chemistry, polymerization is a process of reacting monomer molecules together in a chemical reaction to form three-dimensional networks or polymer chains. There are many forms of polymerization and different systems exist to categorize them. In chemical compounds, polymerization occurs via a variety of reaction mechanisms that vary in complexity due to functional groups present in reacting compounds and their inherent steric effects explained by VSEPR Theory. In more straightforward polymerization, alkenes, which are relatively stable due to σ bonding between carbon atoms form polymers through relatively simple radical reactions; in contrast, more complex reactions such as those that involve substitution at the carbonyl group require more complex synthesis due to the way in which reacting molecules polymerize.

As alkenes can be formed in somewhat straightforward reaction mechanisms, they form useful compounds such as polyethylene and polyvinyl chloride (PVC) when undergoing radical reactions, which are produced in high tonnages each year due to their usefulness in manufacturing processes of commercial products, such as piping, insulation and packaging. Polymers such as PVC are generally referred to as "homopolymers" as they consist of repeated long chains or structures of the same monomer unit, whereas polymers that consist of more than one molecule are referred to as copolymers (or co-polymers).

Other monomer units, such as formaldehyde hydrates or simple aldehydes, are able to polymerize themselves at quite low temperatures (>-80°C) to form trimers; molecules consisting of 3 monomer units which can cyclize to form ring cyclic structures, or undergo further reactions to form tetramers, or 4 monomer-unit compounds. Further compounds either being referred to as oligomers in smaller molecules. Generally, because formaldehyde is an exceptionally reactive electrophile it allows nucleophillic addition of hemiacetal intermediates, which are generally short lived and relatively unstable "mid stage" compounds which react with other molecules present to form more stable polymeric compounds.

Polymerization that is not sufficiently moderated and proceeds at a fast rate can be very hazardous. This phenomenon is known as Hazardous polymerization and can cause fires and explosions.

Step-growth

Step-growth polymers are defined as polymers formed by the stepwise reaction between functional groups of monomers. Most step-growth polymers are also classified as condensation polymers, but not all step-growth polymers (like polyurethanes formed from isocyanate and alcohol bifunctional monomers) release condensates, in this case we talk about addition polymers. Step-growth polymers increase in molecular weight at a very slow rate at lower conversions and reach moderately high molecular weights only at very high conversion (*i.e.* >95%).

To alleviate inconsistencies in these naming methods, adjusted definitions for condensation and addition polymers have been developed. A condensation polymer is defined as a polymer that involves loss of small molecules during its synthesis, or contains functional groups as part of its backbone chain, or its repeat unit does not contain all the atoms present in the hypothetical monomer to which it can be degraded.

Chain-growth

Chain-growth polymerization (or addition polymerization) involves the linking together of molecules incorporating double or triple chemical bonds. These unsaturated *monomers* (the identical molecules that make up the polymers) have extra internal bonds that are able to break and link up with other monomers to form the repeating chain. Chain-growth polymerization is involved in the manufacture of polymers such as polyethylene, polypropylene, and polyvinyl chloride (PVC). A special case of chain-growth polymerization leads to living polymerization.

In the radical polymerization of ethylene, its pi bond is broken, and the two electrons rearrange to create a new propagating centre like the one that attacked it. The form this propagating centre takes depends on the specific type of addition mechanism. There are several mechanisms through which this can be initiated. The free radical mechanism was one of the first methods to be used. Free radicals are very reactive atoms or molecules that have unpaired electrons. Taking the polymerization of ethylene as an example, the free radical mechanism can be divided in to three stages: chain initiation, chain propagation, and chain termination.

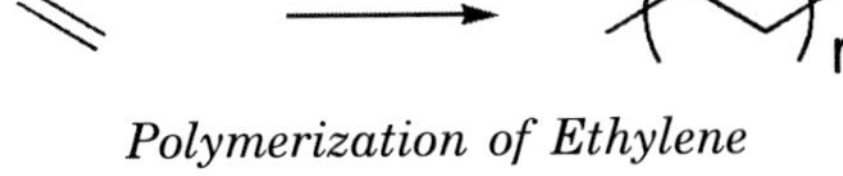

Polymerization of Ethylene

Free radical addition polymerization of ethylene must take place at high temperatures and pressures, approximately 300°C and 2000 atm. While most other free radical polymerizations do not require such extreme temperatures and pressures, they do tend to lack control. One effect of this lack of control is a high degree of branching. Also, as termination occurs randomly, when two chains collide, it is impossible to control the length of individual chains. A newer method of polymerization similar to free radical, but allowing more control involves the Ziegler-Natta catalyst, especially with respect to polymer branching.

Other forms of chain growth polymerization include cationic addition polymerization and anionic addition polymerization. While not used to a large extent in industry yet due to stringent reaction conditions such as lack of water and oxygen, these methods provide ways to polymerize some monomers that cannot be polymerized by free radical methods such as polypropylene. Cationic and anionic mechanisms are also more ideally suited for living polymerizations, although free radical living polymerizations have also been developed.

Ethanol

Ethanol, also called ethyl alcohol, pure alcohol, grain alcohol, or drinking alcohol, is a volatile, flammable, colorless liquid. It is a psychoactive drug and one of the oldest recreational drugs. Best known as the type of alcohol found in alcoholic beverages, it is also used in thermometers, as a solvent, and as a fuel. In common usage, it is often referred to simply as alcohol or spirits.

Ethanol is a straight-chain alcohol, and its molecular formula is C_2H_5OH. Its empirical formula is C_2H_6O. An alternative notation is $CH_3–CH_2–OH$, which indicates that the carbon of a methyl group ($CH_3–$) is attached to the carbon of a methylene group ($–CH_2–$), which is attached to the oxygen of a hydroxyl group (–OH). It is a constitutional isomer of dimethyl ether. Ethanol is often abbreviated as EtOH, using the common organic chemistry notation of representing the ethyl group (C_2H_5) with Et.

The fermentation of sugar into ethanol is one of the earliest organic reactions employed by humanity. The intoxicating effects of ethanol consumption have been known since ancient times. In modern times, ethanol intended for industrial use is also produced from ethylene. Ethanol has widespread use as a solvent of substances intended for human contact or consumption, including scents, flavorings, colorings, and medicines. In chemistry, it is both an essential solvent and a feedstock for the synthesis of other products. It has a long history as a fuel for heat and light, and more recently as a fuel for internal combustion engines.

Ethanol is the systematic name defined by the IUPAC nomenclature of organic chemistry for a molecule with two carbon atoms (prefix "eth-"), having a single bond between them (suffix "-ane"), and an attached -OH group (suffix "-ol").

History

Ethanol has been used by humans since prehistory as the intoxicating ingredient of alcoholic beverages. Dried residue on 9,000-year-old pottery found in China imply that Neolithic people consumed alcoholic beverages.

Although distillation was well known by the early Greeks and Arabs, the first recorded production of alcohol from distilled wine was by the School of Salerno alchemists in the 12th century. The first to mention absolute alcohol, in contrast with alcohol-water mixtures, was Raymond Lull.

In 1796, Johann Tobias Lowitz obtained pure ethanol by filtering distilled ethanol through activated charcoal. Antoine Lavoisier described ethanol as a compound of carbon, hydrogen, and oxygen, and in 1808 Nicolas-Théodore de Saussure determined ethanol's chemical formula. Fifty years later, Archibald Scott Couper published the structural formula of ethanol. It is one of the first structural formulas determined.

Ethanol was first prepared synthetically in 1826 through the independent efforts of Henry Hennel in Great Britain and S.G. Sérullas in France. In 1828, Michael Faraday prepared ethanol by acid-catalysed hydration of ethylene, a process similar to current industrial ethanol synthesis.

Ethanol was used as lamp fuel in the United States as early as 1840, but a tax levied on industrial alcohol during the Civil War made this use uneconomical. The tax was repealed in 1906. Original Ford Model T automobiles ran on ethanol until 1908. With the advent of Prohibition in 1920, ethanol fuel sellers were accused of being allied with moonshiners, and ethanol fuel fell into disuse until late in the 20th century.

Physical Properties

Ethanol is a volatile, colorless liquid that has a slight odor. It burns with a smokeless blue flame that is not always visible in normal light.

The physical properties of ethanol stem primarily from the presence of its hydroxyl group and the shortness of its carbon chain. Ethanol's hydroxyl group is able to participate in hydrogen bonding, rendering it more viscous and less volatile than less polar organic compounds of similar molecular weight.

Ethanol is a versatile solvent, miscible with water and with many organic solvents, including acetic acid, acetone, benzene, carbon tetrachloride, chloroform, diethyl ether, ethylene glycol, glycerol, nitromethane, pyridine, and toluene. It is also miscible with light aliphatic hydrocarbons, such as pentane and hexane, and with aliphatic chlorides such as trichloroethane and tetrachloroethylene.

Ethanol's miscibility with water contrasts with that of longer-chain alcohols (five or more carbon atoms), whose water miscibility decreases sharply as the number of carbons increases. The miscibility of ethanol with alkanes is limited to alkanes up to undecane, mixtures with dodecane and higher alkanes show a miscibility gap below a certain temperature (about 13 °C for dodecane). The miscibility gap tends to get wider with higher alkanes and the temperature for complete miscibility increases.

Ethanol-water mixtures have less volume than the sum of their individual components at the given fractions. Mixing equal volumes of ethanol and water results in only 1.92 volumes of mixture. Mixing ethanol and water is exothermic. At 298 K, up to 777 J/mol are set free.

Mixtures of ethanol and water form an azeotrope at about 89 mole-% ethanol and 11 mole-% water or a mixture of about 96 volume percent ethanol and 4% water at normal pressure and T = 351 K. This azeotropic composition is strongly temperature- and pressure-dependent and vanishes at temperatures below 303 K.

Hydrogen bonding causes pure ethanol to be hygroscopic to the extent that it readily absorbs water from the air. The polar nature of the hydroxyl group causes ethanol to dissolve many ionic compounds, notably sodium and potassium hydroxides, magnesium chloride, calcium chloride, ammonium chloride, ammonium bromide, and sodium bromide. Sodium and potassium chlorides are slightly soluble in ethanol. Because the ethanol molecule also has a nonpolar end, it will also dissolve nonpolar substances, including most essential oils and numerous flavouring, coloring, and medicinal agents.

The addition of even a few percent of ethanol to water sharply reduces the surface tension of water. This property partially explains the "tears of wine" phenomenon. When wine is swirled in a glass, ethanol evaporates quickly from the thin film of wine on the wall of the glass. As the wine's ethanol content decreases, its surface tension increases and the thin film "beads up" and runs down the glass in channels rather than as a smooth sheet.

Mixtures of ethanol and water that contain more than about 50% ethanol are flammable and easily ignited. Alcoholic proof is a widely used measure of how much ethanol (i.e., alcohol) such a mixture contains. In the 18th century, proof was determined by adding a liquor (such as rum) to gunpowder. If the gunpowder still burned, that was considered to be "100 degrees proof" that it was "good" liquor — hence it was called "100 degrees proof".

Ethanol-water solutions that contain less than 50% ethanol may also be flammable if the solution is first heated. Some cooking methods call for wine to be added to a hot pan, causing it to flash boil into a vapour, which is then ignited to burn off excess alcohol.

Ethanol is slightly more refractive than water, having a refractive index of 1.36242 (at ë=589.3 nm and 18.35 °C).

Production

Ethanol is produced both as a petrochemical, through the hydration of ethylene, and biologically, by fermenting sugars with yeast. Which process is more economical depends on prevailing prices of petroleum and grain feed stocks.

Ethylene Hydration

Ethanol for use as an industrial feedstock or solvent (sometimes referred to as synthetic ethanol) is often made from petrochemical feed stocks, primarily by the acid-catalysed hydration of ethylene, represented by the chemical equation

$$C_2H_{4(g)} + H_2O_{(g)} \rightarrow CH_3CH_2OH_{(l)}.$$

The catalyst is most commonly phosphoric acid, adsorbed onto a porous support such as silica gel or diatomaceous earth. This catalyst was first used for large-scale ethanol production by the Shell Oil Company in 1947. The reaction is carried out with an excess of high pressure steam at 300 °C. In the U.S., this process was used on an industrial scale by Union Carbide Corporation and others; but now only LyondellBasell uses it commercially.

In an older process, first practiced on the industrial scale in 1930 by Union Carbide, but now almost entirely obsolete, ethylene was hydrated indirectly by reacting it with concentrated sulfuric acid to produce ethyl sulfate, which was hydrolysed to yield ethanol and regenerate the sulfuric acid:

$$C_2H_4 + H_2SO_4 \rightarrow CH_3CH_2SO_4H$$

$$CH_3CH_2SO_4H + H_2O \rightarrow CH_3CH_2OH + H_2SO_4$$

Fermentation

Ethanol for use in alcoholic beverages, and the vast majority of ethanol for use as fuel, is produced by fermentation. When certain species of yeast (e.g., *Saccharomyces cerevisiae*) metabolize sugar they produce ethanol and carbon dioxide. The chemical equations below summarize the conversion:

$$C_6H_{12}O_6 \rightarrow 2\ CH_3CH_2OH + 2\ CO_2$$

$$C_{12}H_{22}O_{11} + H_2O \rightarrow 4\ CH_3CH_2OH + 4\ CO_2$$

The process of culturing yeast under conditions to produce alcohol is called fermentation. This process is carried out at around 35–40 °C. Toxicity of ethanol to yeast limits the ethanol concentration obtainable by brewing; higher concentrations, therefore, are usually obtained by fortification or distillation. The most ethanol-tolerant strains of yeast can survive up to approximately 15% ethanol by volume.

To produce ethanol from starchy materials such as cereal grains, the starch must first be converted into sugars. In brewing beer, this has traditionally been accomplished by allowing the grain to germinate, or malt, which produces the enzyme amylase. When the malted grain is mashed, the amylase converts the remaining starches into sugars. For fuel ethanol, the hydrolysis of starch into glucose can be accomplished more rapidly by treatment with dilute sulfuric acid, fungally produced amylase, or some combination of the two.

Cellulosic Ethanol

Sugars for ethanol fermentation can be obtained from cellulose. Until recently, however, the cost of the cellulase enzymes capable of hydrolyzing cellulose has been prohibitive. The Canadian firm Iogen brought the first cellulose-based ethanol plant on-stream in 2004. Its primary consumer so far has been the Canadian government, which, along with the United States Department of Energy, has invested heavily in the commercialization of cellulosic ethanol. Deployment of this technology could turn a number of cellulose-containing agricultural by-products, such as corncobs, straw, and sawdust, into renewable energy resources. Other enzyme companies are developing genetically engineered fungi that produce large volumes of cellulase, xylanase, and hemicellulase enzymes. These would convert agricultural residues such as corn stover, wheat straw, and sugar cane bagasse and energy crops such as switchgrass into fermentable sugars.

Cellulose-bearing materials typically also contain other polysaccharides, including hemicellulose. When undergoing hydrolysis, hemicellulose decomposes into mostly five-carbon sugars such as xylose. *S. cerevisiae*, the yeast most commonly used for ethanol production, cannot metabolize xylose. Other yeasts and bacteria are under investigation to ferment xylose and other pentoses into ethanol.

On January 14, 2008, General Motors announced a partnership with Coskata, Inc. The goal is to produce cellulosic ethanol cheaply, with an eventual goal of US$1 per US gallon ($0.30/L) for the fuel. The partnership plans to begin producing the fuel in large quantity by the end of 2008. In June 2009, this goal is still ahead of the firm. By 2011 a full-scale plant will come on line, capable of producing 50 million US gallons (190,000 m^3) to 100 million US gallons (380,000 m^3) of ethanol a year (200–400 ML/a).

Prospective Technologies

The anaerobic bacterium *Clostridium ljungdahlii*, discovered in commercial chicken wastes, can produce ethanol from single-carbon sources including synthesis gas, a mixture of carbon monoxide and hydrogen that can be generated from the partial combustion of either fossil fuels or biomass. Use of these bacteria to produce ethanol from synthesis gas has progressed to the pilot plant stage at the BRI Energy facility in Fayetteville, Arkansas. The BRI technology has been purchased by INEOS.

Another prospective technology is the closed-loop ethanol plant. Ethanol produced from corn has a number of critics who suggest that it is primarily just recycled fossil fuels because of the energy required to grow the grain and convert it into ethanol. There is also the issue of competition with use of corn for food production. However, the closed-loop ethanol plant attempts to address this criticism. In a closed-loop plant, renewable energy for distillation comes from fermented manure, produced from cattle that have been fed the DDSG by-products from grain ethanol production. The concentrated compost nutrients from manure are then used to fertilize the soil and grow the next crop of grain to start the cycle again. Such a process is expected to lower the fossil fuel consumption used during conversion to ethanol by 75%.

An alternative technology allows for the production of biodiesel from distillers grain as an additional value product. Though in an early stage of research, there is some development of alternative production methods that use feed stocks such as municipal waste or recycled products, rice hulls, sugarcane bagasse, small diameter trees, wood chips, and switchgrass.

Testing

Breweries and biofuel plants employ two methods for measuring ethanol concentration. Infrared ethanol sensors measure the vibrational frequency of dissolved ethanol using the CH band at 2900 cm^{-1}. This method uses a relatively inexpensive solid state sensor that compares the CH band with a reference band to calculate the ethanol content. The calculation makes use of the Beer-Lambert law. Alternatively, by measuring the density of the starting material and the density of the product, using a hydrometer, the change in specific gravity during fermentation indicates the alcohol content. This inexpensive and indirect method has a long history in the beer brewing industry.

Purification

Ethylene hydration or brewing produces an ethanol–water mixture. For most industrial and fuel uses, the ethanol must be purified. Fractional distillation can concentrate ethanol to 95.6% by volume (89.5 mole%). This mixture is an azeotrope with a boiling point of 78.1 °C, and cannot be further purified by distillation.

Common methods for obtaining absolute ethanol include desiccation using adsorbents such as starch, corn grits, or zeolites, which adsorb water preferentially, as well as azeotropic distillation and extractive distillation. Most ethanol fuel refineries use an adsorbent or zeolite to desiccate the ethanol stream.

In another method to obtain absolute alcohol, a small quantity of benzene is added to rectified spirit and the mixture is then distilled. Absolute alcohol is obtained in the third fraction, which distills over at 78.3 °C (351.4 K). Because a small amount of the benzene used remains in the solution, absolute alcohol produced by this method is not suitable for consumption, as benzene is carcinogenic.

There is also an absolute alcohol production process by desiccation using glycerol. Alcohol produced by this method is known as spectroscopic alcohol—so called because the absence of benzene makes it suitable as a solvent in spectroscopy.

Grades of Ethanol

Denatured Alcohol

Pure ethanol and alcoholic beverages are heavily taxed, but ethanol has many uses that do not involve consumption by humans. To relieve the tax burden on these uses, most jurisdictions waive the tax when an agent has been added to the ethanol to render it unfit to drink. These include bittering agents such as denatonium benzoate and toxins such as methanol, naphtha, and pyridine. Products of this kind are called *denatured alcohol.*

Absolute Ethanol

Absolute or anhydrous alcohol refers to ethanol with a low water content. There are various grades with maximum water contents ranging from 1% to ppm levels. Absolute alcohol is not intended for human consumption. If azeotropic distillation is used to remove water, it will contain trace amounts of the material separation agent (e.g. benzene).

Absolute ethanol is used as a solvent for laboratory and industrial applications, where water will react with other chemicals, and as fuel alcohol. Spectroscopic ethanol is an absolute ethanol with a low absorbance in ultraviolet and visible light, fit for use as a solvent in ultraviolet-visible spectroscopy.

Pure ethanol is classed as 200 proof in the USA, equivalent to 175 degrees proof in the UK system.

Rectified Spirits

Rectified spirit, an azeotropic composition containing 4% water, is used instead of anhydrous ethanol for various purposes. Wine spirits are about 188 proof. The impurities are different from those in 190 proof laboratory ethanol.

Reactions

Ethanol is classified as a primary alcohol, meaning that the carbon its hydroxyl group attaches to has at least two hydrogen atoms attached to it as well. Many ethanol reactions occur at its hydroxyl group.

Ester Formation

In the presence of acid catalysts, ethanol reacts with carboxylic acids to produce ethyl esters and water:

$$RCOOH + HOCH_2CH_3 \rightarrow RCOOCH_2CH_3 + H_2O$$

This reaction, which is conducted on large scale industrially, requires the removal of the water from the reaction mixture as it is formed. Esters react in the presence of an acid or base to give back the alcohol and a salt. This reaction is known as saponification because it is used in the preparation of soap. Ethanol can also form esters with inorganic acids. Diethyl sulfate and triethyl phosphate are prepared by treating ethanol with sulfur trioxide and phosphorus pentoxide respectively. Diethyl sulfate is a useful ethylating agent in organic synthesis. Ethyl nitrite, prepared from the reaction of ethanol with sodium nitrite and sulfuric acid, was formerly a widely used diuretic.

Dehydration

Strong acid desiccants cause the dehydration of ethanol to form diethyl ether and other byproducts. If the dehydration temperature exceeds around 160 °C, ethylene will be the main product. Millions of kilograms of diethyl ether are produced annually using sulfuric acid catalyst:

$$2\ CH_3CH_2OH \rightarrow CH_3CH_2OCH_2CH_3 + H_2O \text{ (on 120 °C)}$$

Combustion

Complete combustion of ethanol forms carbon dioxide and water vapour:

$$C_2H_5OH\ (l) + 3\ O_2\ (g) \rightarrow 2\ CO_2\ (g) + 3\ H_2O\ (g);$$

$$(\Delta H_c = -1371 \text{ kJ/mol}) \text{ specific heat} = 2.44 \text{ kJ/(kg·K)}$$

Acid-base Chemistry

Ethanol is a neutral molecule and the pH of a solution of ethanol in water is nearly 7.00. Ethanol can be quantitatively converted to its conjugate base, the ethoxide ion ($CH_3CH_2O^-$), by reaction with an alkali metal such as sodium:

$$2\ CH_3CH_2OH + 2\ Na \rightarrow 2\ CH_3CH_2ONa + H_2$$

or a very strong base such as sodium hydride:

$$CH_3CH_2OH + NaH \rightarrow CH_3CH_2ONa + H_2$$

The acidity of water and ethanol are nearly the same, as indicated by their pKa of 15.7 and 16 respectively.

Thus, sodium ethoxide and sodium hydroxide exist in an equilbrium that is closely balanced:

$$CH_3CH_2OH + NaOH \rightleftharpoons CH_3CH_2ONa + H_2O$$

Halogenation

Ethanol is not used industrially as a precursor to ethyl halides, but the reactions are illustrative. Ethanol reacts with hydrogen halides to produce ethyl halides such as ethyl chloride and ethyl bromide via an S_N2 reaction:

$$CH_3CH_2OH + HCl \rightarrow CH_3CH_2Cl + H_2O$$

These reactions require a catalyst such as zinc chloride. HBr requires refluxing with a sulfuric acid catalyst. Ethyl halides can, in principle, also be produced by treating ethanol with more specialized halogenating agents, such as thionyl chloride or phosphorus tribromide.

$$CH_3CH_2OH + SOCl_2 \rightarrow CH_3CH_2Cl + SO_2 + HCl$$

Upon treatment with halogens in the presence of base, ethanol gives the corresponding haloform (CHX_3, where X = Cl, Br, I). This conversion is called the haloform reaction. " An intermediate in the reaction with chlorine is the aldehyde called chloral:

$$4\ Cl_2 + CH_3CH_2OH \rightarrow CCl_3CHO + 5\ HCl$$

Oxidation

Ethanol can be oxidized to acetaldehyde and further oxidized to acetic acid, depending on the reagents and conditions. This oxidation is of no importance industrially, but in the human body, these oxidation reactions are catalysed by the enzyme liver alcohol dehydrogenase.

The oxidation product of ethanol, acetic acid, is a nutrient for humans, being a precursor to acetyl CoA, where the acetyl group can be spent as energy or used for biosynthesis.

Uses

As a Fuel

Energy content of some fuels compared with ethanol:

Fuel type	*MJ/L*	*MJ/kg*	*Research octane number*
Dry wood (20% moisture)		~19.5	
Methanol	17.9	19.9	108.7
Ethanol	21.2	26.8	108.6
E85(85% ethanol, 15% gasoline)	25.2	33.2	105
Liquefied natural gas	25.3	~55	
Autogas (LPG)(60% propane + 40% butane)	26.8	50.	
Aviation gasoline(high-octane gasoline, not jet fuel)	33.5	46.8	100/130 (lean/rich)
Gasohol(90% gasoline + 10% ethanol)	33.7	47.1	93/94
Regular gasoline	34.8	44.4	min. 91
Premium gasoline			max. 104
Diesel	38.6	45.4	25
Charcoal, extruded	50	23	

The largest single use of ethanol is as a motor fuel and fuel additive. Brazil has the largest national fuel ethanol industry. Gasoline sold in Brazil contains at least 25% anhydrous ethanol. Hydrous ethanol (about 95% ethanol and 5% water) can be used as fuel in more than 90% of new cars sold in the country. Brazilian ethanol is produced from sugar cane and noted for high carbon sequestration. The US uses Gasohol (max 10% ethanol) and E85 (85% ethanol) ethanol/gasoline mixtures.

Ethanol may also be utilized as a rocket fuel, and is currently in lightweight rocket-powered racing aircraft.

Australian law limits of the use of pure Ethanol sourced from Sugarcane waste to up to 10% in automobiles. It has been recommended that older cars (and vintage cars designed to use a slower burning fuel) have their valves upgraded or replaced.

Ethanol as a fuel reduces harmful tailpipe emissions of carbon monoxide, particulate matter, oxides of nitrogen, and other ozone-forming pollutants. Argonne National Laboratory analyzed the greenhouse gas emissions of many different engine and fuel combinations. Comparing ethanol blends with gasoline alone, they showed reductions of 8% with the biodiesel/petrodiesel blend known as B20, 17% with the conventional E85 ethanol blend, and that using cellulosic ethanol lowers emissions 64%.

Ethanol combustion in an internal combustion engine yields many of the products of incomplete combustion produced by gasoline and significantly larger amounts of formaldehyde and related species such as acetaldehyde. This leads to a significantly larger photochemical reactivity that generates much more ground level ozone. These data have been assembled into The Clean Fuels Report comparison of fuel emissions and show that ethanol exhaust generates 2.14 times as much ozone as does gasoline exhaust. When this is added into the custom *Localised Pollution Index (LPI)* of The Clean Fuels Report the local pollution (pollution that contributes to smog) is 1.7 on a scale where gasoline is 1.0 and higher numbers signify greater pollution. The California Air Resources Board formalized this issue in 2008 by recognizing control standards for formaldehydes as an emissions control group, much like the conventional NOx and Reactive Organic Gases (ROGs).

World production of ethanol in 2006 was 51 gigalitres (1.3×10^{10} US gal), with 69% of the world supply coming from Brazil and the United States. More than 20% of Brazilian cars are able to use 100% ethanol as fuel, which includes ethanol-only engines and flex-fuel engines. Flex-fuel engines in Brazil are able to work with all ethanol, all gasoline or any mixture of both. In the US flex-fuel vehicles can run on 0% to 85% ethanol (15% gasoline) since higher ethanol blends are not yet allowed or efficient. Brazil supports this population of ethanol-burning automobiles with large national infrastructure that produces ethanol from domestically grown sugar cane. Sugar cane not only has a greater concentration of sucrose than corn (by about 30%), but is also much easier to extract. The bagasse generated by the process is not wasted, but is used in power plants as a surprisingly efficient fuel to produce electricity.

The United States fuel ethanol industry is based largely on corn. According to the Renewable Fuels Association, as of October 30, 2007,

131 grain ethanol bio-refineries in the United States have the capacity to produce 7.0 billion US gallons (26,000,000 m^3) of ethanol per year. An additional 72 construction projects underway (in the U.S.) can add 6.4 billion US gallons (24,000,000 m^3) of new capacity in the next 18 months. Over time, it is believed that a material portion of the H"150-billion-US-gallon (570,000,000 m^3) per year market for gasoline will begin to be replaced with fuel ethanol.

One problem with ethanol is its high miscibility with water, which means that it cannot be efficiently shipped through modern pipelines, like liquid hydrocarbons, over long distances. Mechanics also have seen increased cases of damage to small engines, in particular, the carburetor, attributable to the increased water retention by ethanol in fuel.

Alcoholic Beverages

Ethanol is the principal psychoactive constituent in alcoholic beverages, with depressant effects on the central nervous system. It has a complex mode of action and affects multiple systems in the brain, the most notable one being its agonistic action on the GABA receptors. Similar psychoactives include those that also interact with GABA receptors, such as gamma-hydroxybutyric acid (GHB). Ethanol is metabolized by the body as an energy-providing nutrient, as it metabolizes into acetyl CoA, an intermediate common with glucose and fatty acid metabolism that can be used for energy in the citric acid cycle or for biosynthesis.

Alcoholic beverages vary considerably in ethanol content and in foodstuffs they are produced from. Most alcoholic beverages can be broadly classified as fermented beverages, beverages made by the action of yeast on sugary foodstuffs, or distilled beverages, beverages whose preparation involves concentrating the ethanol in fermented beverages by distillation. The ethanol content of a beverage is usually measured in terms of the volume fraction of ethanol in the beverage, expressed either as a percentage or in alcoholic proof units.

Fermented beverages can be broadly classified by the foodstuff they are fermented from. Beers are made from cereal grains or other starchy materials, wines and ciders from fruit juices, and meads from honey. Cultures around the world have made fermented beverages from numerous other foodstuffs, and local and national names for various fermented beverages abound.

Distilled beverages are made by distilling fermented beverages. Broad categories of distilled beverages include whiskeys, distilled from fermented cereal grains; brandies, distilled from fermented fruit juices; and rum, distilled from fermented molasses or sugarcane juice. Vodka and similar neutral grain spirits can be distilled from any fermented material (grain or potatoes are most common); these spirits are so thoroughly distilled that no tastes from the particular starting material remain. Numerous other spirits and liqueurs are prepared by infusing flavors from fruits, herbs, and spices into distilled spirits. A traditional example is gin, which is created by infusing juniper berries into a neutral grain alcohol.

In a few beverages, ethanol is concentrated by means other than distillation. Applejack is traditionally made by freeze distillation, by which water is frozen out of fermented apple cider, leaving a more ethanol-rich liquid behind. Ice beer (also known by the German term *Eisbier* or *Eisbock*) is also freeze-distilled, with beer as the base beverage. Fortified wines are prepared by adding brandy or some other distilled spirit to partially fermented wine. This kills the yeast and conserves some of the sugar in grape juice; such beverages not only are more ethanol-rich but are often sweeter than other wines.

Alcoholic beverages are sometimes used in cooking, not only for their inherent flavors but also because the alcohol dissolves hydrophobic flavor compounds, which water cannot.

Just as industrial ethanol is used as feedstock for the production of industrial acetic acid, alcoholic beverages are made into culinary/household vinegar: Wine and cider vinegar are both named for their respective source alcohols, whereas malt vinegar is derived from beer.

Feedstock

Ethanol is an important industrial ingredient and has widespread use as a base chemical for other organic compounds. These include ethyl halides, ethyl esters, diethyl ether, acetic acid, ethyl amines, and to a lesser extent butadiene.

Antiseptic

Ethanol is used in medical wipes and in most common antibacterial hand sanitizer gels at a concentration of about 62% v/v as an antiseptic. Ethanol kills organisms by denaturing their proteins and dissolving their lipids and is effective against most bacteria and fungi, and many viruses, but is ineffective against bacterial spores.

Treatment for Poisoning by other Alcohols

Ethanol is sometimes used to treat poisoning by other, more toxic alcohols, in particular methanol and ethylene glycol. Ethanol competes with other alcohols for the alcohol dehydrogenase enzyme, lessening metabolism into toxic aldehyde and carboxylic acid derivatives, and reducing one of the more serious toxic effect of the glycols to crystallize in the kidneys.

Solvent

Ethanol is miscible with water and is a good general purpose solvent. It is found in paints, tinctures, markers, and personal care products such as perfumes and deodorants. It may also be used as a solvent in cooking, such as in vodka sauce.

Historical Uses

Before the development of modern medicines, ethanol was used for a variety of medical purposes. It has been known to be used as a truth drug (as hinted at by the maxim *"in vino veritas"*), as medicine for depression and as an anesthetic.

Ethanol was commonly used as fuel in early bipropellant rocket (liquid propelled) vehicles, in conjunction with an oxidizer such as liquid oxygen. The German V-2 rocket of World War II, credited with beginning the space age, used ethanol, mixed with 25% of water to reduce the combustion chamber temperature. The V-2's design team helped develop U.S. rockets following World War II, including the ethanol-fuelled Redstone rocket, which launched the first U.S. satellite. Alcohols fell into general disuse as more efficient rocket fuels were developed.

Pharmacology

The removal of ethanol through oxidation by alcohol dehydrogenase in the liver from the human body is limited. Hence, the removal of a large concentration of alcohol from blood may follow zero-order kinetics. This means that alcohol leaves the body at a constant rate, rather than having an elimination half-life.

Also, the rate-limiting steps for one substance may be in common with other substances. For instance, the blood alcohol concentration can be used to modify the biochemistry of methanol and ethylene glycol. In this way the oxidation of methanol to the toxic formaldehyde and formic acid in the (human body) can be prevented by giving an appropriate amount of ethanol to a person having ingested methanol. Note that methanol is very toxic and causes blindness and death. A person having ingested ethylene glycol can be treated in the same way.

Drug Effects

Pure ethanol will irritate the skin and eyes. Nausea, vomiting and intoxication are symptoms of ingestion. Long-term use by ingestion can result in serious liver damage. Atmospheric concentrations above one in a thousand are above the European Union Occupational exposure limits.

Short-term

BAC (g/L)	*BAC (%v/v)*	*Symptoms*
0.5	0.05%	Euphoria, talkativeness, relaxation
1	0.1 %	Central nervous system depression, nausea, possible vomiting, impaired motor and sensory function, impaired cognition
>1.4	>0.14%	Decreased blood flow to brain
3	0.3%	Stupefaction, possible unconsciousness
4	0.4%	Possible death
>5.5	>0.55%	Death

Effects on the Central Nervous System

Ethanol is a central nervous system depressant and has significant psychoactive effects in sublethal doses; for specifics, see "Effects of alcohol on the body by dose". Based on its abilities to change the human consciousness, ethanol is considered a psychoactive drug. Death from ethyl alcohol consumption is possible when blood alcohol level reaches 0.4%. A blood level of 0.5% or more is commonly fatal. Levels of even less than 0.1% can cause intoxication, with unconsciousness often occurring at 0.3–0.4%.

The amount of ethanol in the body is typically quantified by blood alcohol content (BAC), which is here taken as weight of ethanol per unit volume of blood. Small doses of ethanol, in general, produce euphoria and relaxation; people experiencing these symptoms tend to become talkative and less inhibited, and may exhibit poor judgment. At higher dosages (BAC > 1 g/L), ethanol acts as a central nervous system depressant, producing at progressively higher dosages, impaired sensory and motor function, slowed cognition, stupefaction, unconsciousness, and possible death.

Ethanol acts in the central nervous system by binding to the GABA-A receptor, increasing the effects of the inhibitory neurotransmitter GABA (i.e., it is a positive allosteric modulator).

Prolonged heavy consumption of alcohol can cause significant permanent damage to the brain and other organs.

In USA, about half of the deaths in car accidents occur in alcohol-related crashes. The risk of a fatal car accident increases exponentially with the level of alcohol in the driver's blood. Most drunk driving laws governing the acceptable levels in the blood while driving or operating heavy machinery set typical upper limits of blood alcohol content (BAC) between 0.05% and 0.08%.

Discontinuing consumption of alcohol after several years of heavy drinking can also be fatal. Alcohol withdrawal can cause anxiety, autonomic dysfunction, seizures, and hallucinations. Delirium tremens is a condition that requires people with a long history of heavy drinking to undertake an alcohol detoxification regimen.

Effects on Metabolism

Ethanol within the human body is converted into acetaldehyde by alcohol dehydrogenase and then into the acetyl in acetyl CoA by acetaldehyde dehydrogenase. Acetyl CoA is the final product of both carbohydrate and fat metabolism, where the acetyl can be further used to produce energy or for biosynthesis. As such, ethanol is a nutrient.

However, the product of the first step of this breakdown, acetaldehyde, is more toxic than ethanol. Acetaldehyde is linked to most of the clinical effects of alcohol. It has been shown to increase the risk of developing cirrhosis of the liver, multiple forms of cancer, and alcoholism.

Drug Interactions

Ethanol can intensify the sedation caused by other central nervous system depressant drugs such as barbiturates, benzodiazepines, opioids, phenothiazines, and anti-depressants.

Magnitude of Effects

Some individuals have less effective forms of one or both of the metabolizing enzymes, and can experience more severe symptoms from ethanol consumption than others. However, those having acquired alcohol tolerance have a greater quantity of these enzymes, and metabolize ethanol more rapidly.

Birth Defects

Ethanol is classified as a teratogen.

Other Effects

Frequent drinking of alcoholic beverages has been shown to be a major contributing factor in cases of elevated blood levels of triglycerides.

Ethanol is not a carcinogen. However, the first metabolic product of ethanol, acetaldehyde, is toxic, mutagenic, and carcinogenic.

Natural Occurrence

Ethanol is a byproduct of the metabolic process of yeast. As such ethanol will be present in any yeast habitat. Ethanol can commonly be found in overripe fruit. Ethanol produced by symbiotic yeast can be found in Bertam Palm blossoms. Although some species such as the Pentailed Treeshrew exhibit ethanol seeking behaviours, most show no interest or avoidance of food sources containing ethanol. Ethanol is also produced during the germination of many plants as a result of natural anerobiosis. Ethanol has been detected in outer space, forming an icy coating around dust grains in interstellar clouds.

3

Hydration Reaction

In organic chemistry, a hydration reaction is a chemical reaction in which a hydroxyl group (OH^-) and a hydrogen cation (an acidic proton) are added to the two carbon atoms bonded together in the carbon-carbon double bond which makes up an alkene functional group. The reaction usually runs in a strong acidic, aqueous solution. Hydration differs from hydrolysis in that hydrolysis cleaves the non-water component in two. Hydration leaves the non-water component intact.

The general chemical equation of the reaction is the following:

$RRC{=}CH_2$ in H_2O/acid $\rightarrow$ $RRC(\text{-}OH)\text{-}CH_3$

In the first step, the acidic proton bonds to the less substituted carbon of the double bond following Markovnikov's rule. In the second step an H_2O molecule bonds to the other, more highly substituted carbon.

The oxygen atom at this point has three bonds and carries a positive charge. Another water molecule comes along and takes up the extra proton.

When carried out in the laboratory, this reaction tends to yield many undesirable side products and in its simple form described here is not considered very useful for the production of alcohol.

Conceptually similar reactions include:

- hydroamination
- hydroalkoxylation
- hydrosilylation
- hydrohalogenation.

Mechanism

This is an example reaction mechanism of the hydration of 1-methylcyclohexene to 1-methylcyclohexanol.

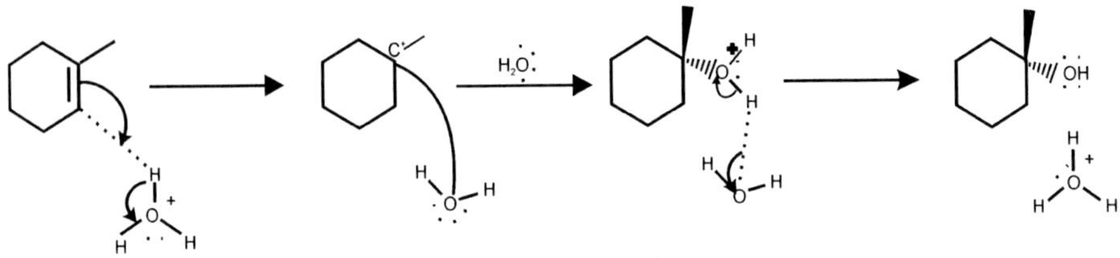

Ethylene Oxide

Ethylene oxide, also called oxirane, is the organic compound with the formula C_2H_4O. It is a cyclic ether. This means that it is composed of two alkyl groups attached to an oxygen atom in a cyclic shape (circular). This colorless flammable gas with a faintly sweet odor is the simplest epoxide, a three-membered ring consisting of two carbon and one oxygen atom. Because of its special molecular structure, ethylene oxide easily participates in the addition reaction, opening its cycle, and thus easily polymerizes. Ethylene oxide is isomeric with acetaldehyde.

Although it is a vital raw material with diverse applications, including the manufacture of products like polysorbate-20 and polyethylene glycol that are often more effective and less toxic than alternative materials, ethylene oxide itself is a very hazardous substance: at room temperature it is a flammable, carcinogenic, mutagenic, irritating, and anaesthetic gas with a misleadingly pleasant aroma.

The chemical reactivity that is responsible for many of ethylene oxide's hazards has also made it a key industrial chemical that supports the living standards of advanced societies. Although too dangerous for direct household use and generally unfamiliar to consumers, ethylene oxide is used industrially for making many consumer products as well as non-consumer chemicals and intermediates. Ethylene oxide is important or critical to the production of detergents, thickeners, solvents, plastics, and various organic chemicals such as ethylene glycol, ethanolamines, simple and complex glycols, polyglycol ethers and other compounds.

As a poison gas that leaves no residue on items it contacts, pure ethylene oxide is a disinfectant that is widely used in hospitals and the medical equipment industry to replace steam in the sterilization of heat-sensitive tools and equipment, such as disposable plastic syringes. Ethylene oxide is industrially produced by direct oxidation of ethylene in the presence of silver catalyst. It is extremely flammable and explosive and is used as a main component of thermobaric weapons; therefore, it is commonly handled and shipped as a refrigerated liquid.

History

Ethylene oxide was first reported in 1859 by the French chemist Charles-Adolphe Wurtz, who prepared it by treating 2-chloroethanol with potassium hydroxide:

$$Cl{-}CH_2CH_2{-}OH + KOH \rightarrow (CH_2CH_2)O + KCl + H_2O$$

Wurtz measured the boiling point of ethylene oxide as 13.5 °C, slightly higher than the present value, and discovered the ability of ethylene oxide to react with acids and salts of metals. Wurtz mistakenly assumed that ethylene oxide has the properties of an organic base. This misconception persisted until 1896 when Georg Bredig found that ethylene oxide is not an electrolyte. Its distinct difference with ethers, in particular, its propensity to join the addition reactions typical of unsaturated compounds, had long been a matter of debate. Only in 1893, the heterocyclic triangular structure of ethylene oxide had been proposed.

The first synthesis method had long remained the only, despite numerous attempts of scientists, including Wurtz himself, to produce ethylene oxide directly from ethylene. Only in 1931, French chemist Theodore Lefort developed a method of direct oxidation of ethylene in the presence of silver catalyst.

Since 1940, almost all industrial production of ethylene oxide has used this process. Sterilization by ethylene oxide for the preservation of spices was patented in 1938 by the American chemist Lloyd Hall. Ethylene oxide achieved industrial importance during World War I as a precursor to both the coolant ethylene glycol and the chemical weapon mustard gas.

Molecular Structure and Properties

The epoxy cycle of ethylene oxide is an almost regular triangle with bond angles of about 60° and a significant angular stress corresponding to the energy of 105 kJ/mol. For comparison, in alcohols the C–O–H angle is about 110°; in ethers, the C–O–C angle is 120°.

The moment of inertia about the principal axes are I_A = 32.921× 10^{-40} g·cm², I_B = 37.926×10^{-40} g·cm² and I_C = 59.510×10^{-40} g·cm². The dipole moment at a temperature in the range 17–176 °C is 6.26× 10^{-30} C·m.

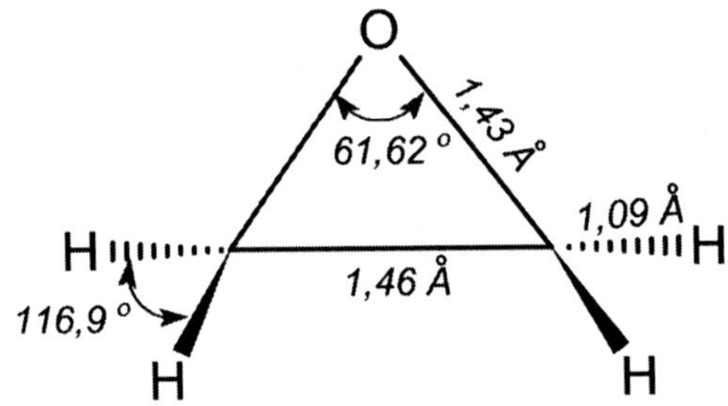

The relative instability of the carbon-oxygen bonds in the molecule is revealed by the comparison in the table of the energy required to break two C–O bonds in the ethylene oxide or one C–O bond in ethanol and dimethyl ether:

Reaction	***ΔH°$_{298}$, kJ/mol***	***Method***
$(C_2H_4)O \rightarrow C_2H_4 + O$ (cleavage of two bonds)	354.38	Calculated, from atomic enthalpies
$C_2H_5OH \rightarrow C_2H_5 + OH$ (breaking one bond)	405.85	Electron impact
$CH_3OCH_3 \rightarrow CH_3O + CH_3$ (breaking one bond)	334.72	Calculated using enthalpies of radicals formation

This instability determines the chemical activity of ethylene oxide and explains the ease of opening its cycle in addition reactions.

Physical Properties

Ethylene oxide is a colorless gas at 25 °C and is a mobile liquid at 0 °C – viscosity of liquid ethylene oxide at 0 °C is about 5.5 times lower than that of water. The gas has a characteristic sweet odor of ether, noticeable when its concentration in air exceeds 500 ppm. Ethylene oxide is readily soluble in water, ethanol, diethyl ether and many organic solvents. Main thermodynamical constants are:

- Standard molar heat capacity, $C_p°$ = 48.19 J/(mol·K);
- Standard enthalpy of formation, $DH°_{298}$ = –51.037 kJ/mol;
- Standard entropy, $S°_{298}$ = 243.4 J/(mol·K);
- Gibbs free energy, $DG°_{298}$ = –11.68 kJ/mol;
- Heat of combustion, $DH_c°$ = –1306 kJ/mol.

The surface tension of liquid ethylene oxide, at the interface with its own steam, is 35.8 mJ/m^2 at –50.1 °C and 27.6 mJ/m^2 at –0.1 °C. The boiling point increases with the vapour pressure as follows: 57.7 (2 atm), 83.6 (5 atm) and 114.0 (10 atm). Viscosity decreases with temperature with the values of 0.577 kPa·s at –49.8 °C, 0.488 kPa·s at –38.2 °C, 0.394 kPa·s at –21.0 °C and 0.320 kPa·s at 0 °C. Between –91 °C and 10.5 °C, vapour pressure p (in mmHg) varies with temperature (T in °C) as $\lg p = 6.251 - 1115.1/(244.14 + T)$.

Table 1 : Properties of Liquid Ethylene Oxide

Temperature, °C	***Steam pressure, kPa***	***Enthalpy of the liquid, J/g***	***Enthalpy of vaporization, J/g***	***Density, kg/L***	***Heat capacity, J/(kg·K)***	***Thermal conductivity, W/(m·K)***
–40 °C	8.35	0	628.6	0.9488	1878	0.20
–20 °C	25.73	38.8	605.4	0.9232	1912	0.18
0 °C	65.82	77.3	581.7	0.8969	1954	0.16
20 °C	145.8	115.3	557.3	0.8697	2008	0.15
40 °C	288.4	153.2	532.1	0.8413	2092	0.14
60 °C	521.2	191.8	505.7	0.8108	2247	0.14
80 °C	875.4	232.6	477.4	0.7794	2426	0.14
100 °C	1385.4	277.8	445.5	0.7443	2782	0.13
120 °C	2088	330.4	407.5	0.7052	3293	N/A*
140 °C	3020	393.5	359.4	0.6609	4225	N/A
160 °C	4224	469.2	297.1	0.608	N/A	N/A
180 °C	5741	551.2	222.5	0.533	N/A	N/A
195.8 °C	7191	N/A	N/A	N/A	N/A	N/A

*N/A – data not available.

Table 2 : Properties of ethylene oxide vapour

Temperature, K	***Entropy, J/(mol·K)***	***Heat of formation, kJ/mol***	***Free energy of formation, kJ/mol***	***Viscosity Pa·s W/(m·K)***	***Thermal conductivity, J/(mol·K)***	***Heat capacity,***
298	242.4	–52.63	–13.10	N/A	N/A	48.28
300	242.8	–52.72	–12.84	9.0	0.012	48.53
400	258.7	–56.53	1.05	13.5	0.025	61.71
500	274.0	–59.62	15.82	15.4	0.038	75.44
600	288.8	–62.13	31.13	18.2	0.056	86.27
700	302.8	–64.10	46.86	20.9	0.075	95.31
800	316.0	–65.61	62.80	N/A	0.090	102.9

*N/A – data not available.

Chemical Properties

Ethylene oxide readily reacts with various compounds, breaking a C–O bond and opening the cycle.

Its typical reactions are with nucleophiles which proceed via the S_N2 mechanism both in acidic (weak nucleophiles: water, alcohols) and alkaline media (strong nucleophiles: OH^-, RO^-, NH_3, RNH_2, RR'NH, etc.). The general reaction scheme is and more specific reactions are described below.

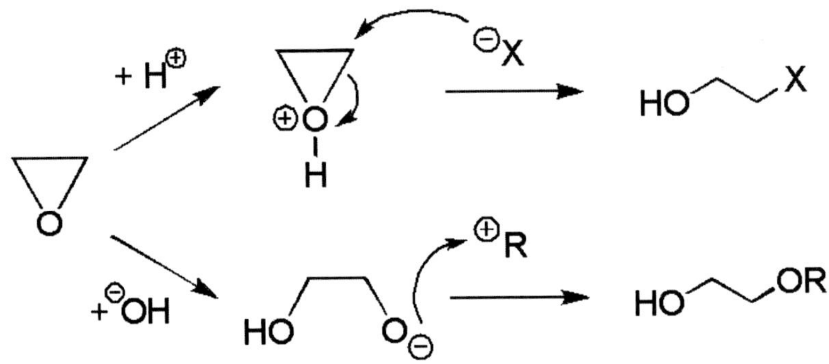

Addition of Water and Alcohols

Aqueous solutions of ethylene oxide are rather stable and can exist for a long time without any noticeable chemical reaction, but adding a small amount of acid, such as strongly diluted sulfuric acid, immediately leads to the formation of ethylene glycol, even at room temperature:

$(CH_2CH_2)O + H_2O \rightarrow HO–CH_2CH_2–OH$

The reaction also occurs in the gas phase, in the presence of a phosphoric acid salt as a catalyst.

The reaction is usually carried out at about 60 °C with a large excess of water, in order to prevent the reaction of the formed ethylene glycol with ethylene oxide that would form di- and triethylene glycol:

$2\ (CH_2CH_2)O + H_2O \rightarrow HO–CH_2CH_2–O–CH_2CH_2–OH$

$3\ (CH_2CH_2)O + H_2O \rightarrow HO–CH_2CH_2–O–CH_2CH_2–O–CH_2CH_2–OH$

The use of alkaline catalysts may lead to the formation of polyethylene glycol:

$n\ (CH_2CH_2)O + H_2O \rightarrow HO–(–CH_2CH_2–O–)_n–H$

Reactions with alcohols proceed similarly yielding ethylene glycol ethers:

$$(CH_2CH_2)O + C_2H_5OH \rightarrow HO\text{–}CH_2CH_2\text{–}OC_2H_5$$

$$2\ (CH_2CH_2)O + C_2H_5OH \rightarrow HO\text{–}CH_2CH_2\text{–}O\text{–}CH_2CH_2\text{–}OC_2H_5$$

Reactions with lower alcohols occur less actively than with water and require more severe conditions, such as heating to 160 °C and pressurizing to 3 MPa and adding an acid or alkali catalyst.

Reactions of ethylene oxide with fatty alcohols proceed in the presence of sodium metal, sodium hydroxide or boron trifluoride and are used for the synthesis of surfactants.

Addition of Carboxylic Acids and their Derivatives

Reactions of ethylene oxide with carboxylic acids in the presence of a catalyst results in incomplete and with anhydrides in complete glycol esters:

$$(CH_2CH_2)O + CH_3COOH \rightarrow HO\text{–}CH_2CH_2\text{–}OCOCH_3$$

$$(CH_2CH_2)O + (CH_3CO)_2O \rightarrow CH_3COO\text{–}CH_2CH_2\text{–}OCOCH_3$$

The addition of acid amides proceeds similarly:

$$(CH_2CH_2)O + CH_3CONH_2 \rightarrow HO\text{–}CH_2CH_2\text{–}NHCOCH_3$$

Addition of ethylene oxide to higher carboxylic acids is carried out at elevated temperatures (typically 140–180 °C) and pressure (0.3–0.5 MPa) in an inert atmosphere, in presence of an alkaline catalyst (concentration 0.01–2%), such as hydroxide or carbonate of sodium or potassium. The carboxylate ion acts as nucleophile in the reaction:

$$RCOOH + OH^- \rightarrow RCOO^- + H_2O$$

$$(CH_2CH_2)O + RCOO^- \rightarrow RCOOCH_2CH_2O^-$$

$$RCOOCH_2CH_2O^- + RCOOH \rightarrow RCOOCH_2CH_2OH + RCOO^-$$

Adding Ammonia and Amines

Ethylene oxide reacts with ammonia forming a mixture of mono-, di- and triethanolamine. The reaction is stimulated by adding a small amount of water.

$$(CH_2CH_2)O + NH_3 \rightarrow HO\text{–}CH_2CH_2\text{–}NH_2$$

$$2\ (CH_2CH_2)O + NH_3 \rightarrow (HO\text{–}CH_2CH_2)_2NH$$

$$3\ (CH_2CH_2)O + NH_3 \rightarrow (HO\text{–}CH_2CH_2)_3N$$

Similarly proceed the reactions with primary and secondary amines:

$$(CH_2CH_2)O + RNH_2 \rightarrow HO\text{–}CH_2CH_2\text{–}NHR$$

Dialkylamino ethanols can further react with ethylene oxide, forming amino polyethylene glycols:

$$n\,(CH_2CH_2)O + R_2NCH_2CH_2OH \rightarrow R_2NCH_2CH_2O\text{–}(\text{–}CH_2CH_2O\text{–})_n\text{–}H$$

Trimethylamine reacts with ethylene oxide in the presence of water, forming choline:

$$(CH_2CH_2)O + (CH_3)_3N + H_2O \rightarrow [HOCH_2CH_2N\,(CH_3)_3]^+OH^-$$

Aromatic primary and secondary amines also react with ethylene oxide, forming the corresponding arylamino alcohols.

Halide Addition

Ethylene oxide readily reacts with aqueous solutions of hydrochloric, hydrobromic and hydroiodic acids to form halohydrins. The reaction occurs easier with the last two acids:

$$(CH_2CH_2)O + HCl \rightarrow HO\text{–}CH_2CH_2\text{–}Cl$$

The reaction with these acids competes with the acid-catalysed hydration of ethylene oxide; therefore, there is always a by-product of ethylene glycol with an admixture of diethylene glycol. For a cleaner product, the reaction is conducted in the gas phase or in an organic solvent. Ethylene fluorohydrin is obtained differently, by boiling hydrogen fluoride with a 5–6% solution of ethylene oxide in diethyl ether. The ether normally has a water content of 1.5–2%; in absence of water, ethylene oxide polymerizes.

Halohydrins can also be obtained by passing ethylene oxide through aqueous solutions of metal halides:

$$2\,(CH_2CH_2)O + CuCl_2 + 2\,H_2O \rightarrow 2\,HO\text{–}CH_2CH_2\text{–}Cl + Cu(OH)_2\downarrow$$

Metalorganic Addition

Interaction of ethylene oxide with organomagnesium compounds, which are Grignard reagents, can be regarded as nucleophilic substitution influenced by carbanion organometallic compounds. The final product of the reaction is a primary alcohol:

$$(CH_2CH_2)O + RMgBr \rightarrow R\text{-}CH_2CH_2\text{-}OMgBr \xrightarrow{H_2O} R\text{-}CH_2CH_2\text{-}OH$$

Similar mechanism is valid for other organometallic compounds, such as alkyl lithium:

$$(CH_2CH_2)O + RLi \rightarrow R\text{-}CH_2CH_2\text{-}OLi \xrightarrow{H_2O} R\text{-}CH_2CH_2\text{-}OH$$

Other Addition Reactions

Addition of Hydrogen Cyanide

Ethylene oxide easily reacts with the hydrogen cyanide forming ethylene cyanohydrin:

$$(CH_2CH_2)O + HCN \rightarrow HO\text{–}CH_2CH_2\text{–}CN$$

A slightly chilled (10–20 °C) aqueous solution of calcium cyanide can be used instead of HCN:

$$2\,(CH_2CH_2)O + Ca(CN)_2 + 2\,H_2O \rightarrow 2\,HO\text{–}CH_2CH_2\text{–}CN + Ca(OH)_2$$

Ethylene cyanohydrin easily loses water, producing acrylonitrile:

$$HO\text{–}CH_2CH_2\text{–}CN \rightarrow CH_2{=}CH\text{–}CN + H_2O$$

Addition of Hydrogen Sulfide and Mercaptans

When reacting with the hydrogen sulfide, ethylene oxide forms 2-mercaptoethanol and thiodiglycol, and with alkylmercaptans it produces 2-alkyl mercaptoetanol:

$$(CH_2CH_2)O + H_2S \rightarrow HO\text{–}CH_2CH_2\text{–}HS$$

$$2\,(CH_2CH_2)O + H_2S \rightarrow (HO\text{–}CH_2CH_2)_2S$$

$$(CH_2CH_2)O + RHS \rightarrow HO\text{–}CH_2CH_2\text{–}SR$$

The excess of ethylene oxide with an aqueous solution of hydrogen sulfide leads to the tris-(hydroxyethyl) sulfonyl hydroxide:

$$3\,(CH_2CH_2)O + H_2S \rightarrow [(HO\text{–}CH_2CH_2)_3S^+]OH^-$$

Addition of Nitrous and Nitric Acids

Reaction of ethylene oxide with aqueous solutions of barium nitrite, calcium nitrite, magnesium nitrite, zinc nitrite or sodium nitrite leads to the formation of 2-nitroethanole:

$$2\,(CH_2CH_2)O + Ca(NO_2)_2 + 2\,H_2O \rightarrow 2\,HO\text{–}CH_2CH_2\text{–}NO_2 + Ca(OH)_2$$

With nitric acid, ethylene oxide forms mono- and dinitroglycols:

$$(CH_2CH_2)O + HNO_3 \rightarrow HO\text{-}CH_2CH_2\text{-}ONO_2 \xrightarrow[-H_2O]{+HNO_3} O_2NO\text{-}CH_2CH_2\text{-}ONO_2$$

Reaction with Compounds Containing Active Methylene Groups

In the presence of alcoholates, reactions of ethylene oxide with compounds containing active methylene group leads to the formation of butyrolactones:

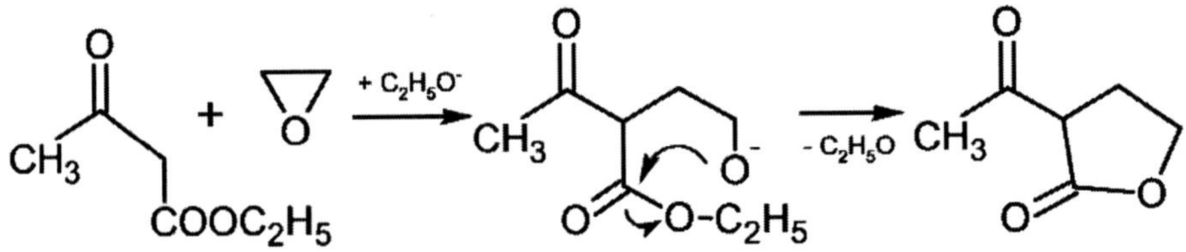

Additions with Aromatic Compounds

Ethylene oxide enters into the Friedel-Crafts reaction with benzene to form phenethyl alcohol:

Styrene can be obtained in one stage if this reaction is conducted at elevated temperatures (315–440 °C) and pressures (0.35–0.7 MPa), in presence of an aluminosilicate catalyst.

Synthesis of Crown Ethers

A series of polynomial heterocyclic compounds, known as crown ethers, can be synthesized with ethylene oxide. One method is the cationic cyclopolymerization of ethylene oxide, limiting the size of the formed cycle:

$$n\ (CH_2CH_2)O \rightarrow (-CH_2CH_2-O-)_n$$

To suppress the formation of other linear polymers the reaction is carried out in a highly dilute solution.

Reaction of ethylene oxide with sulfur dioxide in the presence of caesium salts leads to the formation of an 11-membered heterocyclic compound which has the complexing properties of crown ethers:

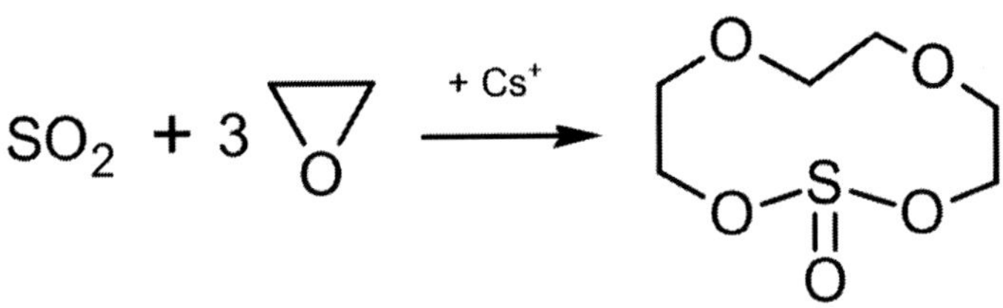

Isomerization

When ethylene oxide is heated to about 400 °C, or to 150–300 °C in the presence of a catalyst (Al_2O_3, H_3PO_4, etc.), it isomerizes into acetaldehyde:

$$(CH_2CH_2)O \xrightarrow{200\ ^\circ C,\ Al_2O_3} CH_3CHO$$

The radical mechanism was proposed by Sidney W. Benson to explain this reaction in the gas phase; it comprises the following stages:

1) $(CH_2CH_2)O \leftrightarrow \bullet CH_2CH_2O\bullet \rightarrow CH_3CHO^*$
2) $CH_3CHO^* \rightarrow CH_3\bullet + CHO\bullet$
3) $CH_3CHO^* + M \rightarrow CH_3CHO + M^*$

In reaction 3), M refers to the wall of the reaction vessel or to a heterogeneous catalyst. The moiety CH_3CHO^* represents a short-lived (lifetime of $10^{-8.5}$ seconds), activated molecule of acetaldehyde. Its excess energy is about 355.6 kJ/mol, which exceeds by 29.3 kJ/mol the binding energy of the C-C bond in acetaldehyde.

In absence of a catalyst, the thermal isomerization of ethylene oxide is never selective and apart from acetaldehyde yields significant amount of by-products.

Reduction Reaction

Ethylene oxide can be hydrogenated into ethanol in the presence of a catalyst, such as nickel, platinum, palladium, boranes, lithium aluminium hydride and some other hydrides.

$$(CH_2CH_2)O + H_2 \xrightarrow{80\ ^\circ C,\ Ni} C_2H_5OH$$

Conversely, with some other catalysts, ethylene oxide may be *reduced* by hydrogen to ethylene with the yield up to 70%. The reduction catalysts include mixtures of zinc dust and acetic acid, of lithium aluminium hydride with titanium trichloride (the reducing agent is actually titanium dichloride, formed by the reaction between $LiAlH_4$ and $TiCl_2$) and of iron(III) chloride with butyllithium in tetrahydrofuran.

$$(CH_2CH_2)O + H_2 \xrightarrow{Zn + CH_3COOH} CH_2{=}CH_2 + H_2O$$

Oxidation

Ethylene oxide can further be oxidized, depending on the conditions, to glycolic acid or carbon dioxide:

$$(CH_2CH_2)O + O_2 \xrightarrow{AgNO_3} HOCH_2COOH$$

Deep gas-phase reactor oxidation of ethylene oxide at 800–1000 K and a pressure of 0.1–1 MPa yields a complex mixture of products containing O_2, H_2, CO, CO_2, CH_4, C_2H_2, C_2H_4, C_2H_6, C_3H_6, C_3H_8 and CH_3CHO.

Dimerization

In the presence of acid catalysts, ethylene oxide can be dimerized into dioxane:

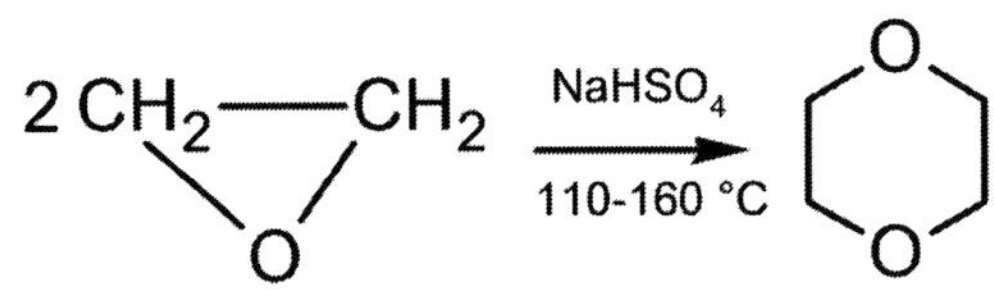

The reaction mechanism is as follows:

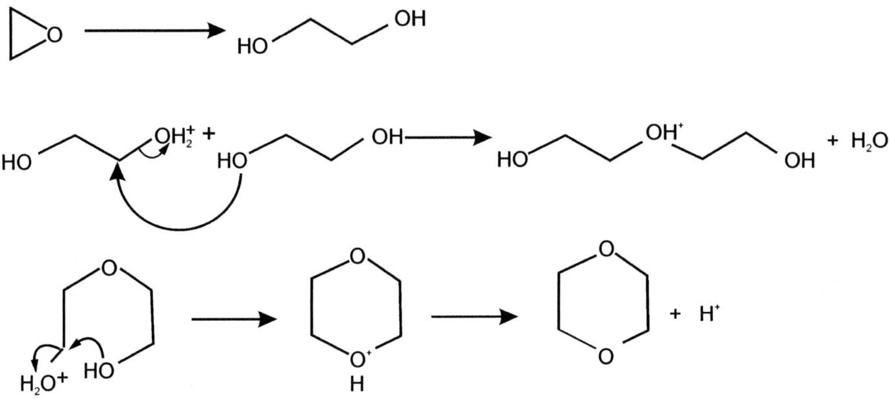

The dimerization reaction is not selective, and there are always by-products, such as acetaldehyde (due to isomerization). The selectivity and speed of dimerization can be increased by adding a catalyst, such as platinum, platinum-palladium or iodine with sulfolan; however, 2-methyl-1,3-dioxolane is formed as a side product in the last case.

Polymerization

Liquid ethylene oxide can form polyethyleneglycols. The polymerization can proceeds via radical and ionic mechanisms, but only the latter has a wide practical application. Cationic polymerization of ethylene oxide is assisted by protonic acids ($HClO_4$, HCl), Lewis acids ($SnCl_4$, BF_3, etc.), organometallic compounds or more complex reagents:

$$n(CH_2CH_2)O \xrightarrow{SnCl_4} (\text{-}CH_2CH_2\text{-}O\text{-})_n$$

The reaction mechanism is as follows. At the first stage, the catalyst (MX_m) is initiated by alkyl-or acylhalogen or by compounds with active hydrogen atoms, usually water, alcohol or glycol:

$$MX_m + ROH \rightarrow MX_mRO^-H^+$$

The resulting active complex reacts with ethylene oxide via the S_N2 mechanism:

$$(CH_2CH_2)O + MX_mRO^-H^+ \rightarrow (CH_2CH_2)O\bullet\bullet\bullet H^+O^-RMX_m$$

$$(CH_2CH_2)O\bullet\bullet\bullet H^+\ O^-RMX_m \rightarrow HO\text{–}CH_2CH_2^+ + MX_mRO^-_2$$

$$HO\text{–}CH_2CH_2^+ + n\ (CH_2CH_2)O \rightarrow HO\text{–}CH_2CH_2\text{–}(O\text{–}CH_2CH_2)_n^+$$

The chain breaks as

$$HO\text{–}CH_2CH_2\text{–}(O\text{–}CH_2CH_2)_n^+ + MX_mRO^- \rightarrow HO\text{–}CH_2CH_2\text{–}(O\text{–}CH_2CH_2)_n\text{–}OR + MX_m$$

$$H(O\text{–}CH_2CH_2)_n\text{–}O\text{–}CH_2\text{–}CH_2^+ + MX_mRO^- \rightarrow H(O\text{–}CH_2CH_2)_n\text{–}O\text{–}CH{=}CH_2 + MX_m + ROH$$

Anionic polymerization of ethylene oxide is assisted by bases, such as alcoholates hydroxides, carbonates or other compounds of alkali or alkaline earth metals. The reaction mechanism is as follows:

$$(CH_2CH_2)O + RONa \rightarrow RO\text{–}CH_2CH_2\text{–}O^-Na^+$$

$$RO\text{–}CH_2CH_2\text{–}O^-Na^+ + n\ (CH_2CH_2)O \rightarrow RO\text{–}(CH_2CH_2\text{–}O)_n\text{–}CH_2CH_2\text{–}O^-Na^+$$

$$RO\text{–}(CH_2CH_2\text{–}O)_n\text{–}CH_2CH_2\text{–}O^-Na^+ \rightarrow RO\text{–}(CH_2CH_2\text{–}O)_n\text{–}CH{=}CH_2 + NaOH$$

$$RO\text{–}(CH_2CH_2\text{–}O)_n\text{–}CH_2CH_2\text{–}O^-Na^+ + H_2O \rightarrow RO\text{–}(CH_2CH_2\text{–}O)_{(n+1)}OH + NaOH$$

Thermal Decomposition

Ethylene oxide is relatively stable to heating – in the absence of a catalyst, it does not dissociate up to 300 °C, and only above 570 °C there is a major exothermic decomposition, which proceeds through the radical mechanism. The first stage involves isomerization, however high temperature accelerated the radical processes. They result in a gas mixture containing acetaldehyde, ethane, ethyl, methane, hydrogen, carbon dioxide, ketene and formaldehyde. High-temperature pyrolysis (830–1200 K) at elevated pressure in an inert atmosphere leads to a more complex composition of the gas mixture, which also contains acetylene and propane. Contrary to the isomerization, initiation of the chain occurs mainly as follows:

$$(CH_2CH_2)O \rightarrow \bullet CH_2CH_2O\bullet \rightarrow CH_2O + CH_2:$$

When carrying the thermal decomposition of ethylene oxide in the presence of transition metal compounds as catalysts, it is possible not only to reduce its temperature, but also to have ethyl as the main product, that is to reverse the ethylene oxide synthesis reaction.

Other Reactions

Thiocyanate ions or thiourea transform ethylene oxide into thiiranes (ethylene sulfides):

$$(CH_2CH_2)O + (NH_2)_2C{=}S \rightarrow (CH_2CH_2)S + (NH_2)_2C{=}O$$

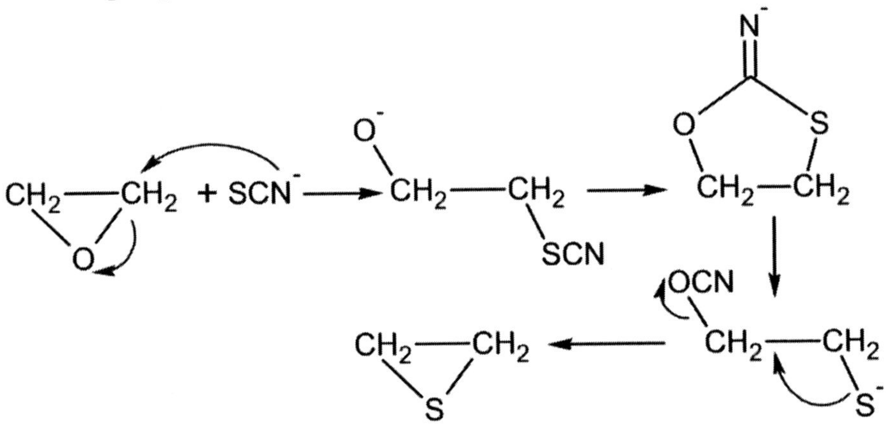

Reaction of phosphorus pentachloride with ethylene oxide produces ethylene dichloride:

$$(CH_2CH_2)O + PCl_5 \rightarrow Cl\text{–}CH_2CH_2\text{–}Cl + POCl_3$$

Other dichloro derivatives of ethylene oxide can be obtained by combined action of sulfuryl chloride ($SOCl_2$) and pyridine and of triphenylphosphine and carbon tetrachloride.

Phosphorus trichloride reacts with ethylene oxide forming chloroethyl esters of phosphorous acid:

$$(CH_2CH_2)O + PCl_3 \rightarrow Cl\text{–}CH_2CH_2\text{–}OPCl_2$$

$$2\ (CH_2CH_2)O + PCl_3 \rightarrow (Cl\text{–}CH_2CH_2\text{–}O)_2PCl$$

$$3\ (CH_2CH_2)O + PCl_3 \rightarrow Cl\text{–}CH_2CH_2\text{–}O)_3P$$

The reaction product of ethylene oxide with acyl chlorides in the presence of sodium iodide is a complex iodoethyl ether:

$$(CH_2CH_2)O + RCOCl + NaI \rightarrow RC(O){-}OCH_2CH_2{-}I + NaCl$$

Heating ethylene oxide to 100 °C with carbon dioxide, in a non-polar solvent in the presence of *bis*-(triphenylphosphine)-nickel(0) results in ethylene carbonate:

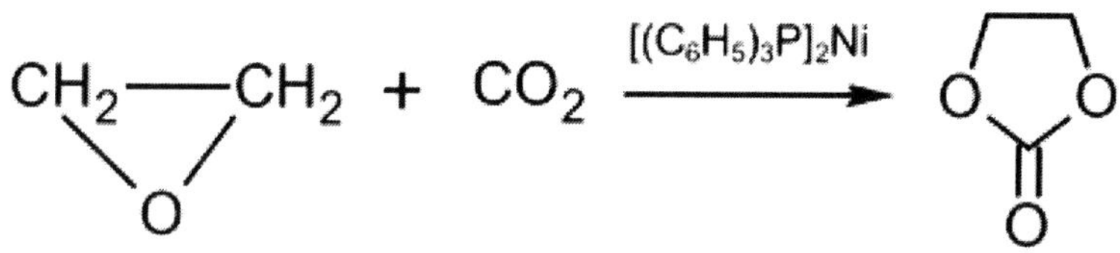

In industry, a similar reaction is carried out at high pressure and temperature in the presence of quaternary ammonium or phosphonium salts as a catalyst.

Reaction of ethylene oxide with formaldehyde at 80–150 °C in the presence of a catalyst leads to the formation of 1,3-dioxolane:

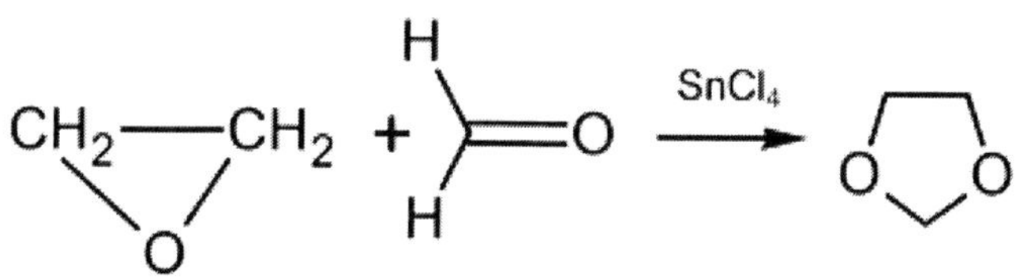

Substituting formaldehyde by other aldehydes or ketones results in a 2-substituted 1,3-dioxolane (yield: 70–85%, catalyst: tetraetilammoniybromid).

Catalytic hydroformylation of ethylene oxide results in hydroxypropanal and further in propane-1,3-diol:

$$(CH_2CH_2)O + CO + H_2 \rightarrow CHO{-}CH_2CH_2{-}OH \xrightarrow{+H_2} HO{-}CH_2CH_2CH_2{-}OH$$

Laboratory Synthesis

Dehydrochlorination of Ethylene and its Derivatives

Dehydrochlorination of 2-chloroethanol, developed by Wurtz back in 1859, still remains one of the most common laboratory methods of producing ethylene oxide:

$$Cl{-}CH_2CH_2{-}OH + NaOH \rightarrow (CH_2CH_2)O + NaCl + H_2O$$

The reaction is carried out at elevated temperature, and beside sodium hydroxide or potassium hydroxide, calcium hydroxide, barium hydroxide, magnesium hydroxide or carbonates of alkali or alkaline earth metals can be used.

Chloroethanol, in turn, is synthesized using one of the following methods:

- By reacting ethylene glycol with hydrochloric acid:

$HO–CH_2CH_2–OH + HCl \rightarrow HO–CH_2CH_2–Cl + H_2O$

- By reacting ethylene with hypochlorous acid:

$CH_2{=}CH_2 + HOCl \rightarrow HO–CH_2CH_2–Cl$

- By chlorination of ethylene:

$CH_2{=}CH_2 + Cl_2 + H_2O \rightarrow HO–CH_2CH_2–Cl + HCl$

Another convenient and old method of ethylene oxide synthesis is reaction of an alkali with chloroethyl acetate:

$$Cl–CH_2CH_2–OCOCH_3 + 2\ KOH \rightarrow (CH_2CH_2)O + KCl + CH_3COOK + H_2O$$

With a high yield (90%) ethylene oxide can be produced by reacting calcium oxide with ethyl hypochlorite; substituting calcium by other alkaline earth metals reduces the reaction yield:

$2\ CH_3CH_2–OCl + CaO \rightarrow 2\ (CH_2CH_2)O + CaCl_2 + H_2O$

In turn, ethylhypochlorite is synthesized as follows:

$Cl_2 + NaOH + CH_3CH_2OH \rightarrow CH_3CH_2OCl + NaCl + H_2O$

Direct Oxidation of Ethylene by Peroxy Acids

Ethylene can be directly oxidized into ethylene oxide using peroxy acids, for example, peroxybenzoic or *meta*-chloro-peroxybenzoic acid:

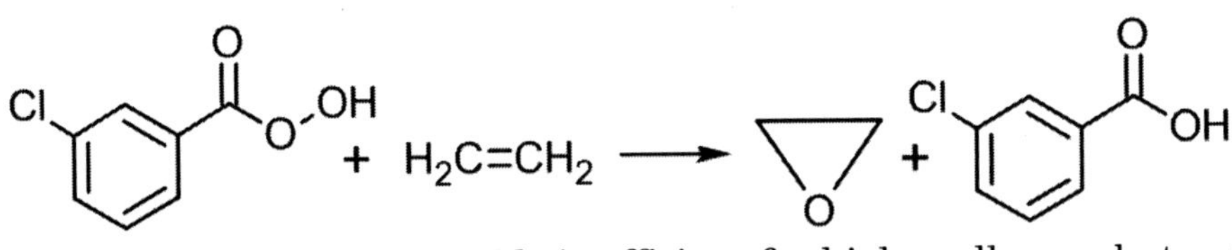

Oxidation by peroxy acids is efficient for higher alkenes, but not for ethylene. The above reaction is slow and has low yield, therefore it is not used in the industry.

Other Preparative Methods

Other synthesis methods include reaction of diiodo ethane with silver oxide:

$I{-}CH_2CH_2{-}I + Ag_2O \rightarrow (CH_2CH_2)O + 2\ AgI$

and decomposition of ethylene carbonate at 200–210 °C in the presence of hexachloroethane:

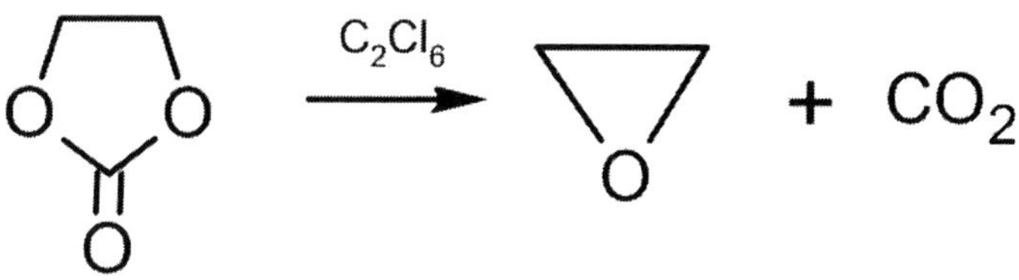

Industrial Synthesis

History

Commercial production of ethylene oxide dates back to 1914 when BASF built the first factory which used the chlorohydrin process (reaction of ethylene chlorohydrin with calcium hydroxide). The chlorohydrin process was unattractive for several reasons, including low efficiency and loss of valuable chlorine into calcium chloride. More efficient direct oxidation of ethylene by air was invented by Lefort in 1931 and in 1937 Union Carbide opened the first plant using this process.

It was further improved in 1958 by Shell Oil Co. by replacing air with oxygen and using elevated temperature of 200–300 °C and pressure (1–3 MPa). This more efficient routine accounted for about half of ethylene oxide production in the 1950s in the U.S., and after 1975 it completely replaced the previous methods.

Chlorohydrin Process of Production of Ethylene Oxide

Although the chlorohydrin process is almost entirely superseded in the industry by the direct oxidation of ethylene, the knowledge of this method is still important for educational reasons and because it is still used in the production of propylene oxide. The process consists of three major steps: synthesis of ethylene chlorohydrin, dehydrochlorination of ethylene chlorohydrin to ethylene oxide and purification of ethylene oxide. Those steps are carried continuously. In the first column, hypochlorination of ethylene is carried out as follows:

$Cl_2 + H_2O \rightarrow HOCl + HCl$

$CH_2{=}CH_2 + HOCl \rightarrow OH{-}CH_2CH_2{-}Cl$

$CH_2{=}CH_2 + Cl_2 \rightarrow Cl{-}CH_2CH_2{-}Cl$

To suppress the conversion of ethylene into the ethylene dichloride (the last reaction), the concentration of ethylene is maintained at about 4–6%, and the solution is heated by steam to the boiling point.

Next, aqueous solution of ethylene chlorohydrin enters the second column, where it reacts with a 30% solution of calcium hydroxide at 100 °C:

$$2\ OH–CH_2CH_2–Cl + Ca(OH)_2 \rightarrow 2\ (CH_2CH_2)O + CaCl_2 + H_2O$$

The produced ethylene oxide is purified by rectification. The chlorohydrin process allows to reach 95% conversion of ethylene chlorohydrin. The yield of ethylene oxide is about 80% of the theoretical value; for 1 ton of ethylene oxide, about 200 kg of ethylene dichloride is produced.

Direct Oxidation of Ethylene

Usage in Global Industry

Direct oxidation of ethylene was patented by Lefort in 1931. This method was repeatedly modified for industrial use, and at least four major variations are known. They all use oxidation by oxygen or air and a silver-based catalyst, but differ in the technological details and hardware implementations.

Union Carbide (currently a division of Dow Chemical Company) was the first company to develop the direct oxidation process. Since 1994, it uses the so-called METEOR process (Most Effective Technology for Ethylene Oxide Reactions) which is characterized by high productivity, low initial capital investment and low operating costs. The method is the exclusive property of the company; it is used only at its own plants and accounts for about 20% of the global ethylene oxide production.

A similar production method was developed by Scientific Design Co., but it received wider use because of the licensing system – it accounts for 25% of the world's production and for 75% of world's licensed production of ethylene oxide. A proprietary variation of this method is used by Japan Catalytic Chemical Co., which adapted synthesis of both ethylene oxide and ethylene glycol in a single industrial complex.

A different modification was developed Shell International Chemicals BV. Their method is rather flexible with regard to the specific requirements of specific industries; it is characterized by high selectivity with respect to the ethylene oxide product and long lifetime of the catalyst (3 years). It accounts for about 40% of global production.

Older factories typically use air for oxidation whereas newer plants and processes, such as METEOR and Japan Catalytic, favour oxygen.

Chemistry and Kinetics of the Direct Oxidation Process

Formally, the direct oxidation process is expressed by the following equation:

$$2CH_2{=}CH_2 + O_2 \xrightarrow{Ag} 2(CH_2CH_2)O$$

However, significant yield of carbon dioxide and water is observed in practice, which can be explained by the complete oxidation of ethylene or ethylene oxide:

$$CH_2{=}CH_2 + 3\ O_2 \rightarrow 2\ CO_2 + 2\ H_2O$$

$$2\ (CH_2CH_2)O + 5\ O_2 \rightarrow 4\ CO_2 + 4\ H_2O$$

The process of heterogeneous catalytic oxidation of ethylene was studied by P. A. Kilty and W. M. H. Sachtler, who suggested the following mechanism:

$$O_2 + 4\ Ag(adj) \rightarrow 4\ Ag + 2\ O^{2-}(ads)$$

$$O_2 + Ag \rightarrow Ag^+ + O_2^-$$

$$O_2^-(ads) + CH_2{=}CH_2 \rightarrow (CH_2CH_2)O + O(ads)$$

$$6\ O\ (ads) + CH_2{=}CH_2 \rightarrow 2\ CO_2 + 2\ H_2O$$

Here (ads) refers to particles adsorbed on the catalyst surface and (adj) to particles of silver, directly adjacent to the oxygen atoms.

Thus the overall reaction is expressed as

$$7\ CH_2{=}CH_2 + 6\ O_2 \rightarrow 6\ (CH_2CH_2)O + 2\ CO_2 + 2\ H_2O$$

and the maximum degree of conversion of ethylene to ethylene oxide is 6/7 or 85.7%.

The catalyst for the reaction is metallic silver deposited on various matrixes, including pumice, silica gel, various silicates and aluminosilicates, alumina and silicon carbide, and activated by certain additives (antimony, bismuth, barium peroxide, etc.). The process temperature was optimized as 220–280 °C. Lower temperatures reduce the activity of the catalyst, and higher temperatures promote the complete oxidation of ethylene thereby reducing the yield of ethylene oxide. Elevated pressure of 1–3 MPa increases the productivity of the catalyst and facilitates absorption of ethylene oxide from the reacting gases.

Whereas oxidation by air is still being used, oxygen (> 95% purity) is preferred for several reasons, such as higher molar yield of ethylene oxide (75–82% for oxygen vs. 63–75% for air), higher reaction rate (no gas dilution) and no need of separating nitrogen in the reaction products.

Applications

Ethylene oxide is one of the most important raw materials used in the large-scale chemical production. Most ethylene oxide is used for synthesis of ethylene glycols, including diethylene glycol and triethylene glycol, that accounts for up to 75% of global consumption. Other important products include ethylene glycol ethers, ethanolamines and ethoxylates. Among glycols, ethylene glycol is used as antifreeze, in the production of polyester and polyethylene terephthalate (PET – raw material for plastic bottles), liquid coolants and solvents. Polyethyleneglycols are used in perfumes, cosmetics, pharmaceuticals, lubricants, paint thinners and plasticizers. Ethylene glycol ethers are part of brake fluids, detergents, solvents, lacquers and paints. Other products of ethylene oxide. Ethanolamines are used in the manufacture of soap and detergents and for purification of natural gas. Ethoxylates are reaction products of ethylene oxide with higher alcohols, acids or amines. They are used in the manufacture of detergents, surfactants, emulsifiers and dispersants.

Whereas synthesis of ethylene glycols is the major application of ethylene oxide, its percentage varies greatly depending on the region: from 44% in the Western Europe, 63% in Japan and 73% in North America to 90% in the rest of Asia and 99% in Africa.

Production of Ethylene Glycol

Ethylene glycol is industrially produced by non-catalytic hydration of ethylene oxide at a temperature of 200 °C and a pressure of 1.5–2 MPa:

$$(CH_2CH_2)O + H_2O \rightarrow HOCH_2CH_2OH$$

By-products of the reaction are diethylene glycol, triethylene glycol and polyglycols with the total of about 10%, which are separated from the ethylene glycol by distillation at reduced pressure. Another synthesis method is the reaction of ethylene oxide and CO_2 (temperature 80–120 °C and pressure of 5.2 MPa) yielding ethylene carbonate and its subsequent hydrolysis with decarboxylation:

$$(CH_2CH_2)O + CO_2 \rightarrow (O\text{-}CH_2CH_2\text{-}O)C{=}O \xrightarrow[-CO_2]{+H_2O} HOCH_2CH_2OH$$

Modern technologies of production of ethylene glycol include the following. Shell OMEGA technology (Only Mono-Ethylene Glycol Advantage) is a two-step synthesis of ethylene carbonate using a phosphonium halide as a catalyst. The glycol yield is 99–99.5%, with other glycols practically absent. The main advantage of the process is production of pure ethylene glycol without the need for further purification. The first commercial plant which uses this method was opened in 2008 in South Korea. Dow METEOR (Most Effective Technology for Ethylene Oxide Reactions) is an integrated technology for producing ethylene oxide and its subsequent hydrolysis into ethylene glycol. The glycol yield is 90–93%. The main advantage of the process is relative simplicity, using fewer stages and less equipment.

Production of Glycol Ethers

The major industrial esters of mono-, di- and triethylene glycols are methyl, ethyl and normal butyl ethers, as well as their acetates and phthalates. The synthesis involves reaction of the appropriate alcohol with ethylene oxide:

$$(CH_2CH_2)O + ROH \rightarrow HOCH_2CH_2OR$$

$$(CH_2CH_2)O + HOCH_2CH_2OR \rightarrow HOCH_2CH_2OCH_2CH_2OR$$

$$(CH_2CH_2)O + HOCH_2CH_2OCH_2CH_2OR \rightarrow HOCH_2CH_2OCH_2CH_2OCH_2CH_2OR$$

The reaction of monoesters with an acid or its anhydride leads to the formation of the esters:

$$CH_3COOH + HOCH_2CH_2OR \rightarrow ROCH_2CH_2OCOCH_3 + H_2O$$

Production of Ethanolamines

In the industry, ethanolamines (mono-, di- and triethanolamines) are produced by reacting ammonia and ethylene oxide in anhydrous medium at a temperature of 40–70 °C and pressure of 1.5–3.5 MPa:

$$CH_2CH_2)O + NH_3 \rightarrow HOCH_2CH_2NH_2$$

$$2\,(CH_2CH_2)O + NH_3 \rightarrow (HOCH_2CH_2)_2NH$$

$$3\,(CH_2CH_2)O + NH_3 \rightarrow (HOCH_2CH_2)_3N$$

All three ethanolamines are produced in the process, while ammonia and part of methylamine are recycled. The final products are separated by vacuum distillation. Hydroxyalkylamines are produced in a similar process:

$$CH_2CH_2)O + RNH_2 \rightarrow HOCH_2CH_2NHR$$

$$2\,(CH_2CH_2)O + RNH_2 \rightarrow (HOCH_2CH_2)_2NR$$

Monosubstituted products are formed by reacting a large excess of amine with ethylene oxide in presence of water and at a temperature below 100 °C. Disubstituted products are obtained with a small excess of ethylene oxide, at a temperature of 120–140 °C and a pressure of 0.3–0.5 MPa.

Production of Ethoxylates

Industrial production of ethoxylates is realized by a direct reaction of higher alcohols, acids or amines with ethylene oxide in the presence of an alkaline catalyst at a temperature of 120–180 °C. Modern plants producing ethoxylates are usually based on the BUSS LOOP reactors technology, which is based on a three-stage continuous process.

In the first stage, the initiator or catalyst of the reaction and the feedstock are fed into the container, where they are mixed, heated and vacuum dried. Then reaction is carried out in a special insulated reactor in an inert atmosphere (nitrogen) to prevent a possible explosion of ethylene oxide. Finally, the reaction mixture is neutralized, degassed and purified.

Production of Acrylonitrile

Currently, most acrylonitrile (90% in 2008) is produced by the SOHIO method, which is based on the catalytic oxidation of propylene in the presence of ammonia and bismuth phosphomolybdate. However, until 1960 a key production process was addition of hydrogen cyanide to ethylene oxide, followed by dehydration of the resulting cyanohydrin::

$$(CH_2CH_2)O + HCN \rightarrow HOCH_2CH_2CN \xrightarrow[-H_2O]{} CH_2{=}CH{-}CN$$

Addition of hydrocyanic acid to ethylene oxide is carried out in the presence of a catalyst (sodium hydroxide and diethylamine), and dehydration of cyanohydrin occurs in the gas phase upon the catalytic action of aluminium oxide.

Other Uses

The direct use of ethylene oxide accounts for only 0.05% (2004 data) of its global production. Ethylene oxide is used as a fumigant and disinfecting agent, as a mixture with carbon dioxide (8.5–80% of ethylene oxide), nitrogen or dichlorodifluoromethane (12% ethylene oxide). It is applied for gas-phase sterilization of medical equipment and instruments, packaging materials and clothing, surgical and scientific equipment; for processing of storage facilities (tobacco, packages of grain, sacks of rice, etc.), clothing, furs and valuable documents.

Ethylene oxide is also used as a flame retardant, accelerator of maturation of tobacco leaves and fungicide. Ethylene oxide is also used as a main component of thermobaric weapons (fuel-air explosives).

Identification of Ethylene Oxide

The simplest qualitative reaction for ethylene oxide uses its property to precipitate insoluble hydroxides of metals when it is passed through aqueous solutions of their salts, for example

$$2\ (CH_2CH_2)O + MnCl_2 + 2\ H_2O \rightarrow 2\ HO{-}CH_2CH_2{-}Cl + Mn(OH)_2\downarrow$$

Similarly, ethylene oxide is detected by the bright pink colour of the indicator when passing air through aqueous solutions of some salts of sodium or potassium (chlorides, iodides, thiosulfates, etc.) with the addition of phenolphthalein:

$$(CH_2CH_2)O + NaCl + H_2O \rightarrow HO{-}CH_2CH_2{-}Cl + NaOH$$

Other methods of ethylene oxide detection are colour reactions with pyridine derivatives and hydrolysis of ethylene glycol with periodic acid. The produced iodic acid is detected with silver nitrate.

The main physical method of ethylene oxide detection is gas chromatography.

Fire and Explosion Hazards

Ethylene oxide is extremely flammable and its mixtures with air are explosive. When heated, it may rapidly expand causing fire and explosion. The autoignition temperature is 429 °C, minimum inflammable content in the air is 2.7%, and the NPFA rating is NFPA 704.

Fires caused by ethylene oxide are extinguished by traditional media, including foam, carbon dioxide or water. Extinguishing of burning ethylene oxide is complicated by that it can continue burning in an inert atmosphere and in water solutions. Fire suppression is reached only upon dilution with water above 22:1.

Physiological Effects

Effect on Microorganisms

Ethylene oxide inhibits growth of microorganisms (disinfectant properties) and when present in high concentrations, can completely destroy them. Strong alkylating properties make ethylene oxide a universal poison for protoplasm: it causes clotting of proteins, deactivation of enzymes and other biologically important components of a living organism.

Ethylene oxide acts more strongly against bacteria, especially gram-positive bacteria, than against yeast and fungi. The disinfectant effect of ethylene oxide is similar to that of sterilization by heat, but because of limited penetration, it affects only the surface.

Effects on Humans and Animals

Ethylene oxide is an alkylating agent; it has irritating, sensitizing and narcotic effects. Chronic exposure to ethylene oxide also induces mutagenic effects. The International Agency for Research on Cancer classifies ethylene oxide into group 1, meaning it is a proven carcinogen. A 2003 study of 7,576 women exposed while at work in commercial sterilization facilities in the U.S. suggests ethylene oxide is associated with breast cancer incidence. A 2004 follow up study analyzing 18,235 men and women workers exposed to ethylene oxide from 1987 to 1998 concluded "There was little evidence of any excess cancer mortality for the cohort as a whole, with the exception of bone cancer based on small numbers. Positive exposure-response trends for lymphoid tumors were found for males only. Reasons for the sex specificity of this effect are not known. There was also some evidence of a positive exposure-response for breast cancer mortality."

Ethylene oxide is toxic by inhalation with an U.S. OSHA permissible exposure limit of 1 ppm calculated as a time weighted average (TWA) over 8 hours of 1 ppm, and a short term exposure limit (excursion limit) calculated as a TWA over 15 minutes of 5 ppm. [29 CFR 19101.1048]. At concentrations in the air about 200 parts per million, ethylene oxide irritates mucous membranes of the nose and throat; higher contents cause damage to the trachea and bronchi, progressing into the partial collapse of the lungs. High concentrations can cause pulmonary edema and damage the cardiovascular system; the damaging effect of ethylene oxide may occur only after 72 hours after exposure. The maximum content of ethylene oxide in the air according to the U.S. standards (ACGIH) is 1.8 mg/m^3. NIOSH has determined that the Immediately Dangerous to Life and Health level (IDLH) is 800 ppm.

Because the odor threshold for ethylene oxide varies between 250 and 700 ppm, the gas will already be at toxic concentrations when it can be smelled. Even then, the odor of ethylene oxide is sweet, aromatic, and can easily be mistaken for the pleasant aroma of diethyl ether, a common laboratory solvent of very low toxicity. In view of these insidious warning properties, continuous electrochemical monitors are standard practice, and it is forbidden to use ethylene oxide to fumigate building interiors in the EU and some other jurisdictions.

Ethylene oxide causes acute poisoning, accompanied by the following symptoms: slight heartbeat, muscle twitching, flushing, headache, diminished hearing, acidosis, vomiting, dizziness, transient loss of consciousness and a sweet taste in the mouth. Acute intoxication is accompanied by a strong throbbing headache, dizziness, difficulty in speech and walking, sleep disturbance, pain in the legs, weakness, stiffness, sweating, increased muscular irritability, transient spasm of retinal vessels, enlargement of the liver and suppression of its antitoxic functions.

Ethylene easily penetrates through the clothing and footwear, causing skin irritation and dermatitis with the formation of blisters, fever and leukocytosis.

The median lethal doses (LD_{50}, or a dose required to kill half the members of a tested population after a certain time) for ethylene oxide are 72 mg/kg (rat, oral) and 187 mg/kg (rat, subcutaneous injection).

Redox

Redox (portmanteau for *red*uction-*ox*idation) reactions describe all chemical reactions in which atoms have their oxidation number changed. This can be either a simple redox process, such as the oxidation of carbon to yield carbon dioxide (CO_2) or the reduction of carbon by hydrogen to yield methane (CH_4), or a complex process such as the oxidation of glucose ($C_6H_{12}O_6$) in the human body through a series of complex electron transfer processes.

Redox reactions, or oxidation-reduction reactions, have a number of similarities to acid-base reactions. Fundamentally, redox reactions are a family of reactions that are concerned with the transfer of electrons between species. Like acid-base reactions, redox reactions are a matched set, that is there cannot be an oxidation reaction without a reduction reaction happening simultaneously. Oxidation refers to the loss of electrons, while reduction refers to the gain of electrons. Each reaction by itself is called a "half-reaction", simply because there must be two half-reactions to form a whole reaction. Thus, in notating redox reactions, chemists typically write out the electrons explicitly:

The term comes from the two concepts of reduction and oxidation. It can be explained in simple terms:

- Oxidation is the *loss* of electrons or an *increase* in oxidation state by a molecule, atom, or ion.
- Reduction is the *gain* of electrons or a *decrease* in oxidation state by a molecule, atom, or ion.

Though sufficient for many purposes, these descriptions are not precisely correct. Oxidation and reduction properly refer to *a change in oxidation number* — the actual transfer of electrons may never occur. Thus, oxidation is better defined as an *increase in oxidation number*, and reduction as a *decrease in oxidation number*. In practice, the transfer of electrons will always cause a change in oxidation number, but there are many reactions that are classed as "redox" even though no electron transfer occurs (such as those involving covalent bonds).

Oxidizing and Reducing Agents

In redox processes, the reductant transfers electrons to the oxidant. Thus, in the reaction, the reductant or *reducing agent* loses electrons and is oxidized, and the oxidant or *oxidizing agent* gains electrons and is reduced. The pair of an oxidizing and reducing agent that are involved in a particular reaction is called a redox pair.

Oxidizers

Substances that have the ability to oxidize other substances are said to be oxidative or oxidizing and are known as oxidizing agents, oxidants, or oxidizers. Put another way, the oxidant (oxidizing agent) removes electrons from another substance; i.e., it oxidizes other substances, and is thus itself reduced. And, because it "accepts" electrons, it is also called an electron acceptor.

Oxidants are usually chemical elements or substances with elements in high oxidation numbers (e.g., H_2O_2, MnO^{-4}, CrO_3, Cr_2O2^{-7}, OsO_4) or highly electronegative substances/elements that can gain one or two extra electrons by oxidizing an element or substance (O, F, Cl, Br).

Reducers

Substances that have the ability to reduce other substances are said to be reductive or reducing and are known as reducing agents, reductants, or reducers. The reductant (reducing agent) transfers electrons to another substance; i.e., it reduces others, and is thus itself oxidized. And, because it "donates" electrons, it is also called an electron donor. Electron donors can also form charge transfer complexes with electron acceptors.

Reductants in chemistry are very diverse. Electropositive elemental metals, such as lithium, sodium, magnesium, iron, zinc, and aluminium, are good reducing agents. These metals donate or *give away* electrons readily. *Hydride transfer reagents*, such as $NaBH_4$ and $LiAlH_4$, are widely used in organic chemistry, primarily in the reduction of carbonyl compounds to alcohols. Another method of reduction involves the use of hydrogen gas (H_2) with a palladium, platinum, or nickel catalyst. These *catalytic reductions* are used primarily in the reduction of carbon-carbon double or triple bonds.

Standard Reduction Potential

Reduction potential is used to calculate the standard electrode potential (E^ocell).

This is the equation most commonly seen in textbooks:

$E^o\text{cell} = E^o_{red} + E^o_{ox}$.

where: E^ocell is the standard electrode potential (in volts).

E^o_{red} is standard reduction potential of the reducing agent.

E^o_{ox} (standard oxidation potential) is negative of the standard reduction potential of the oxidizing agent.

Though the following equation is generally more useful as one is usually given only reduction potentials, not oxidation potentials:

$$E^{o}cell = E^{o}_{red} - E^{o}_{ox}$$

or the equivalent:

$$E^{o}cell = E^{o}_{cathode} - E^{o}_{anode}$$

where:

$E^{o}cell$ is the standard electrode potential (in volts).

E^{o}_{red} ($E^{o}_{cathode}$) is standard reduction potential of the reducing agent.

E^{o}_{ox} (E^{o}_{anode}) is the standard reduction potential of the oxidizing agent.

Examples of Redox Reactions

A good example is the reaction between hydrogen and fluorine in which hydrogen is being oxidized and fluorine is being reduced:

$$H_2 + F_2 \rightarrow 2\ HF$$

We can write this overall reaction as two half-reactions:

the oxidation reaction:

$$H_2 \rightarrow 2\ H^+ + 2\ e^-$$

and the reduction reaction:

$$F_2 + 2\ e^- \rightarrow 2\ F^-$$

Analyzing each half-reaction in isolation can often make the overall chemical process clearer. Because there is no net change in charge during a redox reaction, the number of electrons in excess in the oxidation reaction must equal the number consumed by the reduction reaction (as shown above).

Elements, even in molecular form, always have an oxidation number of zero. In the first half-reaction, hydrogen is oxidized from an oxidation number of zero to an oxidation number of +1. In the second half-reaction, fluorine is reduced from an oxidation number of zero to an oxidation number of –1.

When adding the reactions together the electrons are canceled:

$$H_2 \rightarrow 2\ H^+ + 2\ e^-$$

$$F_2 + 2\ e^- \rightarrow 2\ F^-$$

$$H_2 + F_2 \rightarrow 2\ H^+ + 2\ F^-$$

And the ions combine to form hydrogen fluoride:

$$H_2 + F_2 \rightarrow 2\ H^+ + 2\ F^- \rightarrow 2\ HF$$

Displacement Reactions

Redox occurs in single displacement reactions or substitution reactions. The redox component of these types of reactions is the change of oxidation state (charge) on certain atoms, not the actual exchange of atoms in the compounds. For example, in the reaction between iron and copper(II) sulfate solution:

$$Fe + CuSO_4 \rightarrow FeSO_4 + Cu$$

The ionic equation for this reaction is:

$$Fe + Cu^{2+} \rightarrow Fe^{2+} + Cu$$

As two half-equations, it is seen that the iron is oxidized:

$$Fe \rightarrow Fe^{2+} + 2\ e^-$$

And the copper is reduced:

$$Cu^{2+} + 2\ e^- \rightarrow Cu$$

Other Examples

- The oxidation of iron(II) to iron(III) by hydrogen peroxide in the presence of an acid:

$$Fe^{2+} \rightarrow Fe^{3+} + e^-$$

$$H_2O_2 + 2\ e^- \rightarrow 2\ OH^-$$

Overall equation:

$$2\ Fe^{2+} + H_2O_2 + 2\ H^+ \rightarrow 2\ Fe^{3+} + 2\ H_2O$$

- The reduction of nitrate to nitrogen in the presence of an acid (denitrification):

$$2\ NO_3^- + 10\ e^- + 12\ H^+ \rightarrow N_2 + 6\ H_2O$$

- Oxidation of elemental iron to iron(III) oxide by oxygen (commonly known as rusting, which is similar to tarnishing):

$$4\ Fe + 3\ O_2 \rightarrow 2\ Fe_2O_3$$

- The combustion of hydrocarbons, such as in an internal combustion engine, which produces water, carbon dioxide, some partially oxidized forms such as carbon monoxide, and heat energy. Complete oxidation of materials containing carbon produces carbon dioxide.

- In organic chemistry, the stepwise oxidation of a hydrocarbon by oxygen produces water and, successively, an alcohol, an aldehyde or a ketone, a carboxylic acid, and then a peroxide.

Redox Reactions in Industry

The primary process of reducing ore to produce metals is discussed in the article on Smelting.

Oxidation is used in a wide variety of industries such as in the production of cleaning products and oxidizing ammonia to produce nitric acid, which is used in most fertilizers.

Redox reactions are the foundation of electrochemical cells.

The process of electroplating uses redox reactions to coat objects with a thin layer of a material, as in chrome-plated automotive parts, silver plating cutlery, and gold-plated jewelry.

The production of compact discs depends on a redox reaction, which coats the disc with a thin layer of metal film.

Redox Reactions in Biology

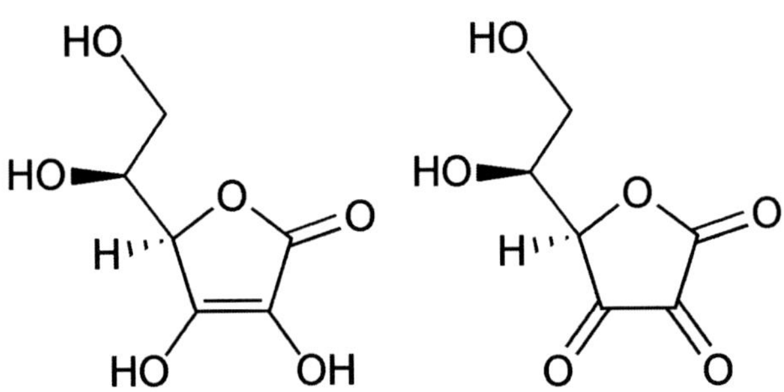

Left: ascorbic acid (reduced form of Vitamin C)

Right: dehydroascorbic acid (oxidized form of Vitamin C)

Many important biological processes involve redox reactions.

Cellular respiration, for instance, is the oxidation of glucose ($C_6H_{12}O_6$) to CO_2 and the reduction of oxygen to water. The summary equation for cell respiration is:

$$C_6H_{12}O_6 + 6\ O_2 \rightarrow 6\ CO_2 + 6\ H_2O$$

The process of cell respiration also depends heavily on the reduction of NAD^+ to NADH and the reverse reaction (the oxidation of NADH to NAD^+). Photosynthesis and Cellular respiration are complementary but photosynthesis is not the reverse of the redox reaction in cell respiration:

$$6\ CO_2 + 6\ H_2O + \text{light energy} \rightarrow C_6H_{12}O_6 + 6\ O_2$$

Biological energy is frequently stored and released by means of redox reactions. Photosynthesis involves the reduction of carbon dioxide into sugars and the oxidation of water into molecular oxygen. The reverse reaction, respiration, oxidizes sugars to produce carbon dioxide and water. As intermediate steps, the reduced carbon compounds are used to reduce nicotinamide adenine dinucleotide (NAD^+), which then contributes to the creation of a proton gradient, which drives the synthesis of adenosine triphosphate (ATP) and is maintained by the reduction of oxygen. In animal cells, mitochondria perform similar functions.

Free radical reactions are redox reactions that occur as a part of homeostasis and killing microorganisms, where an electron detaches from a molecule and then reattaches almost instantaneously. Free radicals are a part of redox molecules and can become harmful to the human body if they do not reattach to the redox molecule or an antioxidant. Unsatisfied free radicals can spur the mutation of cells they encounter and are thus causes of cancer.

The term redox state is often used to describe the balance of NAD^+/NADH and $NADP^+$/NADPH in a biological system such as a cell or organ. The redox state is reflected in the balance of several sets of metabolites (e.g., lactate and pyruvate, beta-hydroxybutyrate and acetoacetate), whose interconversion is dependent on these ratios. An abnormal redox state can develop in a variety of deleterious situations, such as hypoxia, shock, and sepsis. Redox signaling involves the control of cellular processes by redox processes.

Redox proteins and their genes must be co-located for redox regulation according to the CoRR hypothesis for the function of DNA in mitochondria and chloroplasts.

Redox Cycling

A wide variety of aromatic compounds are enzymatically reduced to form free radicals that contain one more electron than their parent compounds. In general, the electron donor is any of a wide variety of flavoenzymes and their coenzymes. Once formed, these anion free radicals reduce molecular oxygen to superoxide, and regenerate the unchanged parent compound. The net reaction is the oxidation of the flavoenzyme's coenzymes and the reduction of molecular oxygen to form superoxide. This catalytic behaviour has been described as futile cycle or redox cycling.

Examples of redox cycling-inducing molecules are the herbicide paraquat and other viologens and quinones such as menadione.

Redox Reactions in Geology

In geology, redox is important to both the formation of minerals, mobilization of minerals, and in some depositional environments. In general, the redox state of most rocks can be seen in the colour of the rock. Red is associated with oxidizing conditions of formation, and green is typically associated with reducing conditions. White (bleached rock) can also be associated with reducing conditions. Famous examples of redox conditions affecting geological processes include uranium deposits and Moqui marbles.

Balancing Redox Reactions

Describing the overall electrochemical reaction for a redox process requires a *balancing* of the component half-reactions for oxidation and reduction. In general, for reactions in aqueous solution, this involves adding H^+, OH^-, H_2O, and electrons to compensate for the oxidation changes.

Acidic Media

In acidic media, H^+ ions and water are added to half reactions to balance the overall reaction.

For example, when manganese(II) reacts with sodium bismuthate:

Unbalanced reaction:

$$Mn^{2+}(aq) + NaBiO_3(s) \rightarrow Bi^{3+}(aq) + MnO_4^- (aq)$$

Oxidation:

$$4\ H_2O(l) + Mn^{2+}(aq) \rightarrow MnO^{-}4(aq) + 8\ H^+(aq) + 5\ e^-$$

Reduction:

$$2\ e^- + 6\ H^+ + BiO^{-3}(s) \rightarrow Bi^{3+}(aq) + 3\ H_2O(l)$$

The reaction is balanced by scaling the two half-cell reactions to involve the same number of electrons (multiplying the oxidation reaction by the number of electrons in the reduction step and vice versa):

$8\ H_2O(l) + 2\ Mn^{2+}(aq) \rightarrow 2\ MnO^-$

$4(aq) + 16\ H^+(aq) + 10\ e^-$

$10\ e^- + 30\ H^+ + 5\ BiO^-$

$3(s) \rightarrow 5\ Bi^{3+}(aq) + 15\ H_2O(l)$

Adding these two reactions eliminates the electrons terms and yields the balanced reaction:

$14\ H^+(aq) + 2\ Mn^{2+}(aq) + 5\ NaBiO_3(s) \rightarrow 7\ H_2O(l) + 2\ MnO^-$

$4(aq) + 5\ Bi^{3+}(aq) + 5\ Na^+(aq)$

Basic Media

In basic media, OH^- ions and water are added to half reactions to balance the overall reaction.

For example, in the reaction between potassium permanganate and sodium sulfite:

Unbalanced reaction:

$$KMnO_4 + Na_2SO_3 + H_2O \rightarrow MnO_2 + Na_2SO_4 + KOH$$

Reduction:

$$3\ e^- + 2\ H_2O + MnO_4^- \rightarrow MnO_2 + 4\ OH^-$$

Oxidation:

$$2\ OH^- + SO_3^{2-} \rightarrow SO_4^{2-} + H_2O + 2\ e^-$$

Balancing the number of electrons in the two half-cell reactions gives:

$$6\ e^- + 4\ H_2O + 2\ MnO_4^- \rightarrow 2\ MnO_2 + 8\ OH^-$$

$$6\ OH^- + 3\ SO_3^{2-} \rightarrow 3\ SO_4^{2-} + 3\ H_2O + 6\ e^-$$

Adding these two half-cell reactions together gives the balanced equation:

$$2\ KMnO_4 + 3\ Na_2SO_3 + H_2O \rightarrow 2\ MnO_2 + 3\ Na_2SO_4 + 2\ KOH$$

Memory Aids

The key terms involved in redox are often confusing to students. For example, an element that is oxidized loses electrons; however, that element is referred to as the reducing agent. Likewise, an element that is reduced gains electrons and is referred to as the oxidizing agent. Acronyms or mnemonics are commonly used to help remember what is happening:

- "LEO the lion says GER" — Loss of Electrons is Oxidation, Gain of Electrons is Reduction.
- "LEORA says GEROA" — Loss of Electrons is Oxidation (Reducing Agent) and Gain of Electrons is Reduced (Oxidizing Agent).

Another commonly used memory aid is:

- "OIL RIG"—Oxidation Is Loss of electrons, Reduction Is Gain of electrons.

Ethylene Glycol

Ethylene glycol (IUPAC name: ethan-1,2-diol) is an organic compound widely used as an automotive antifreeze and a precursor to polymers. In its pure form, it is an odorless, colorless, syrupy, sweet-tasting liquid. Ethylene glycol is toxic, and ingestion can result in death.

Ethylene glycol is not to be confused with diethylene glycol, a heavier ether diol, or with polyethylene glycol, a nontoxic polyether polymer.

Production

Historical Aspects and Natural Occurrence

Ethylene glycol was first prepared in 1859 by the French chemist Charles-Adolphe Wurtz from ethylene glycol diacetate via saponification with potassium hydroxide and, in 1860, from the hydration of ethylene oxide. There appears to have been no commercial manufacture or application of ethylene glycol prior to World War I, when it was synthesized from ethylene dichloride in Germany and used as a substitute for glycerol in the explosives industry.

In the United States, semicommercial production of ethylene glycol via ethylene chlorohydrin started in 1917. The first large-scale commercial glycol plant was erected in 1925 at South Charleston, West Virginia, by Carbide and Carbon Chemicals Co. (now Union Carbide Corp.). By 1929, ethylene glycol was being used by almost all dynamite manufacturers.

In 1937, Carbide started up the first plant based on Lefort's process for vapour-phase oxidation of ethylene to ethylene oxide. Carbide maintained a monopoly on the direct oxidation process until 1953, when the Scientific Design process was commercialized and offered for licenses. This molecule has been observed in outer space.

Current Methods

Ethylene glycol is produced from ethylene (ethene), via the intermediate ethylene oxide. Ethylene oxide reacts with water to produce ethylene glycol according to the chemical equation:

$$C_2H_4O + H_2O \rightarrow HO–CH_2CH_2–OH$$

This reaction can be catalysed by either acids or bases, or can occur at neutral pH under elevated temperatures. The highest yields of ethylene glycol occur at acidic or neutral pH with a large excess of water. Under these conditions, ethylene glycol yields of 90% can be achieved. The major byproducts are the ethylene glycol oligomers diethylene glycol, triethylene glycol, and tetraethylene glycol. About 6.7 billion kilograms are produced annually.

A higher selectivity is achieved by use of the OMEGA-Process. In the OMEGA process, the ethylene oxide is first converted with carbon dioxide (CO_2) to ethylene carbonate to then react with water in a second step to selectively produce mono-ethylene glycol. The carbon dioxide is released in this step again and can be fed back into the process circuit. The carbon dioxide comes in part from the ethylene oxide production, where a part of the ethylene is completely oxidized.

Uses

The major end uses of ethylene glycol are to make PET melt, which accounted for 86% of total ethylene glycol consumption in 2009, and antifreeze, at around 7% of total consumption. Because this material is cheaply available, it finds many niche applications.

Coolant and Heat Transfer Agent

The major use of ethylene glycol is as a medium for convective heat transfer in, for example, automobiles and liquid cooled computers. Ethylene glycol is also commonly used in chilled water air conditioning systems that place either the chiller or air handlers outside, or systems that must cool below the freezing temperature of water. In geothermal heating/cooling systems, ethylene glycol is the fluid that transports heat through the use of a geothermal heat pump. The ethylene glycol either gains energy from the source (lake, ocean, water well) or dissipates heat to the source, depending if the system is being used for heating or cooling.

Pure ethylene glycol has a specific heat capacity about one half that of water. So, while providing freeze protection and an increased boiling point, ethylene glycol lowers the specific heat capacity of water mixtures relative to pure water. A 50/50 mix by mass has a specific heat capacity of about 0.75 BTU/lb F, thus requiring increased flow rates in same system comparisons with water. Additionally, the increase in boiling point over pure water inhibits nucleate boiling on heat transfer surfaces thus reducing heat transfer efficiency in some cases, such as gasoline engine cylinder walls. Therefore, pure ethylene glycol should not be used as an engine coolant in most cases.

Antifreeze

Due to its low freezing point and tendency to form glasses, ethylene glycol resists freezing. A mixture of 60% ethylene glycol and 40% water does not freeze until temperatures below "45 °C ("49°F). Diethyleneglycol behaves similarly. It is used as a deicing fluid for windshields and aircraft. The antifreeze capabilities of ethylene glycol have made it an important component of vitrification (anticrystallization) mixtures for low-temperature preservation of biological tissues and organs.

Table 3 : Ethylene glycol freezing point vs. concentration in water

Weight Percent EG (%)	*Freezing Point (deg F)*	*Freezing Point (deg C)*
0	32	0
10	25	–4
20	20	–7
30	5	–15
40	–10	–23
50	–30	–34
60	–55	–48
70	–60	–51
80	–50	–45
90	–20	–29
100	–10	–12

Ethylene glycol disrupts hydrogen bonding when dissolved in water. Pure ethylene glycol freezes at about "12°C (10.4°F), but when mixed with water molecules, neither can readily form a solid crystal structure, and therefore the freezing point of the mixture is depressed significantly. The minimum freezing point is observed when the ethylene glycol percent in water is about 70%, as shown below. This is the reason pure ethylene glycol is not used as an antifreeze—water is a necessary component as well.

However, the boiling point for aqueous ethylene glycol increases monotonically with increasing ethylene glycol percentage. Thus, the use of ethylene glycol not only depresses the freezing point, but also elevates the boiling point such that the operating range for the heat transfer fluid is broadened on both ends of the temperature scale. The increase in boiling temperature is due to pure ethylene glycol having a much higher boiling point and lower vapour pressure than pure water; there is no chemical stabilization against boiling of the liquid phase at intermediate compositions, as there is against freezing.

Table 4 : Ethylene glycol boiling point vs. concentration in water

Weight Percent EG (%)	***Boiling Point (deg F)***	***Boiling Point (deg C)***
0	212	100
10	215	102
20	215	102
30	220	104
40	220	104
50	225	107
60	230	110
70	240	116
80	255	124
90	285	140
100	387	197

Precursor to Polymers

In the plastics industry, ethylene glycol is important precursor to polyester fibres and resins. Polyethylene terephthalate, used to make plastic bottles for soft drinks, is prepared from ethylene glycol.

Hydrate Inhibition

Because of its high boiling point and affinity for water, ethylene glycol is a useful desiccant. Ethylene glycol is widely used to inhibit the formation of natural gas clathrates (hydrates) in long multiphase pipelines that convey natural gas from remote gas fields to an onshore processing facility. Ethylene glycol can be recovered from the natural gas and reused as an inhibitor after purification treatment that removes water and inorganic salts. Natural gas is dehydrated by ethylene glycol. In this application, ethylene glycol flows down from the top of a tower and meets a rising mixture of water vapour and hydrocarbon gases. Dry gas exits from the top of the tower. The glycol and water are separated, and the glycol recycled. Instead of removing water, ethylene glycol can also be used to depress the temperature at which hydrates are formed. The purity of glycol used for hydrate suppression (monoethylene glycol) is typically around 80%, whereas the purity of glycol used for dehydration (triethylene glycol) is typically 95 to more than 99%. Moreover, the injection rate for hydrate suppression is much lower than the circulation rate in a glycol dehydration tower.

Niche Applications

Minor uses of ethylene glycol include the manufacture of capacitors, as a chemical intermediate in the manufacture of 1,4-dioxane, and as an additive to prevent corrosion in liquid cooling systems for personal computers. Ethylene glycol is also used in the manufacture of some vaccines, but it is not itself present in these injections. It is used as a minor (1–2%) ingredient in shoe polish and also in some inks and dyes. Ethylene glycol has seen some use as a rot and fungal treatment for wood, both as a preventative and a treatment after the fact. It has been used in a few cases to treat partially rotted wooden objects to be displayed in museums. It is one of only a few treatments that are successful in dealing with rot in wooden boats, and is relatively cheap. Ethylene glycol may also be one of the minor ingredients in screen cleaning solutions, along with the main ingredient isopropyl alcohol. Ethylene glycol is commonly used as a preservative for biological specimens, especially in secondary schools during dissection as a safer alternative to formaldehyde. It can also be used in killing jars. It is also used as part of the water-based fluid used to control subsea oil and gas production equipment.

Chemical Reactions

Ethylene glycol is used as a protecting group for carbonyl groups in organic synthesis. Treating a ketone or aldehyde with ethylene glycol in the presence of an acid catalyst (e.g., p-toluenesulfonic acid; $BF_3 \bullet Et_2O$) gives the corresponding a 1,3-dioxolane, which is resistant to bases and other nucleophiles. The 1,3-dioxolane protecting group can thereafter be removed by further acid hydrolysis. In this example, isophorone was protected using ethylene glycol with p-toluenesulfonic acid in moderate yield. Water was removed by azeotropic distillation to shift the equilibrium to the right.

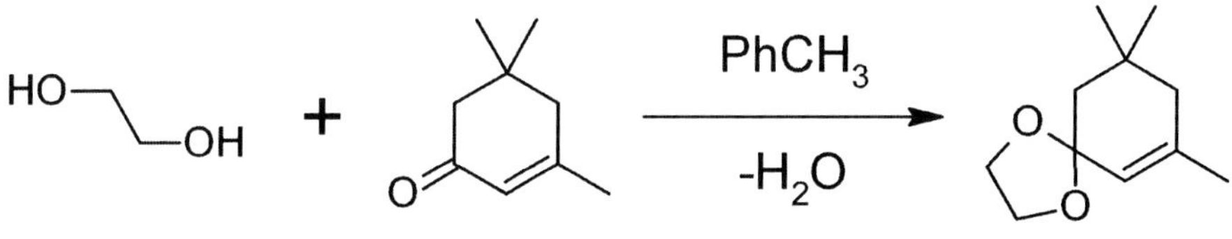

Toxicity

Ethylene glycol is moderately toxic with an oral LDLO = 786 mg/kg for humans. The major danger is due to its sweet taste. Because of that, children and animals are more inclined to consume large quantities of it than of other poisons. Upon ingestion, ethylene glycol is oxidized to glycolic acid which is, in turn, oxidized to oxalic acid, which is toxic. It and its toxic byproducts first affect the central nervous system, then the heart, and finally the kidneys. Ingestion of sufficient amounts can be fatal if untreated.

According to the annual report of the American Association of Poison Control Centers' National Poison Data System in 2007, there were about 1000 total cases resulting in 16 deaths. The 2008 American Association of Poison Control Centers' National Poison Data System annual report lists 7 deaths.(Toxicity, Ethylene Glycol)

In the Environment

- The primary source of ethylene glycol in the environment is from run-off at airports where it is used in deicing agents for runways and airplanes. Ethylene glycol can also enter the environment through the disposal of products that contain it.
- Ethylene glycol in air will break down in about 10 days.
- Ethylene glycol in water and in soil will break down within several days to a few weeks.

Antifreeze

Antifreeze is a freeze preventive used in internal combustion engines and other heat transfer applications, such as HVAC chillers and solar water heaters. The purpose of antifreeze is to prevent a rigid enclosure from undergoing catastrophic deformation due to expansion when water turns to ice. Antifreezes are chemical compounds added to water to reduce the freezing point of the mixture below the lowest temperature that the system is likely to encounter. Either the *additive* or the *mixture* may be referred to as antifreeze.

An antifreeze mixture achieves freezing point depression for a cold environment and also achieves boiling point elevation to enable higher liquid temperatures. This is described as the action of a colligative agent, which can properly be referred to as both antifreeze and "anti-boil" when used for both properties. Careful selection of an antifreeze can enable a wide temperature range in which the mixture remains in the liquid phase, which is critical to efficient heat transfer and the proper functioning of heat exchangers.

Automotive and Internal Combustion Engine Use

Most automotive engines are "water"-cooled to remove waste heat, although the "water" is actually antifreeze/water mixture and not plain water. The term *engine coolant* is widely used in the automotive industry, which covers its primary function of convective heat transfer for internal combustion engines. When used in an automotive context, corrosion inhibitors are added to help protect vehicles' radiators, which often contain a range of electrochemically incompatible metals (aluminium, cast iron, copper, brass, solder, et cetera). Water pump seal lubricant is also added.

Antifreeze was developed to overcome the shortcomings of water as a heat transfer fluid. In some engines freeze plugs are placed in areas of the engine block where coolant flows in order to protect the engine from freeze damage if the ambient temperature drops below the freezing point of the antifreeze/water mixture. These should not be confused with core plugs, whose purpose is to allow removal of sand used in the casting process of engine blocks (core plugs will be pushed out if the coolant freezes, though).

On the other hand, if the engine coolant gets too hot, it might boil while inside the engine, causing voids (pockets of steam), leading to localized hot spots and the catastrophic failure of the engine. If plain water were to be used as an engine coolant, it would promote galvanic corrosion. Proper engine coolant and a pressurized coolant system can help obviate the problems which make plain water incompatible with automotive engines. With proper antifreeze a wide temperature range can be tolerated by the engine coolant, such as –34 °F ("37 °C) to +265 °F (129 °C) for 50% (by volume) propylene glycol diluted with water and a 15 psi pressurized coolant system. Early engine coolant antifreeze was methanol (methyl alcohol), still used in windshield washer fluid. As radiator caps were vented, not sealed, the methanol was lost to evaporation, requiring frequent replenishment to avoid freezing of the coolant. Methanol also accelerates corrosion of the metals, especially aluminium, used in the engine and cooling systems. Ethylene glycol was developed, and soon replaced methanol as an engine cooling system antifreeze. It has a very low volatility compared to methanol and to water.

Other Uses

The most common water-based antifreeze solutions used in electronics cooling are mixtures of water and either ethylene glycol (EGW) or propylene glycol (PGW). The use of ethylene glycol has a longer history, especially in the automotive industry. However, EGW solutions formulated for the automotive industry often have silicate based rust inhibitors that can coat and/or clog heat exchanger surfaces. The use of PGW as a coolant is becoming more common primarily because it is environmentally friendly and non-toxic. Ethylene glycol is listed as a toxic chemical requiring care in handling and disposal.

Ethylene glycol has desirable thermal properties, including a high boiling point, low freezing point, stability over a wide range of temperatures, and high specific heat and thermal conductivity. It also has a low viscosity and, therefore reduced pumping requirements. Although EGW has more desirable physical properties than PGW, the latter coolant is used in applications where toxicity might be a concern. PGW is generally recognized as safe for use in food or food processing applications, and can also be used in enclosed spaces.

Primary Agents

Most antifreeze is made by mixing distilled water with some kind of alcohol.

Methanol

Methanol (also known as methyl alcohol, carbinol, wood alcohol, wood naphtha or wood spirits) is a chemical compound with chemical formula CH_3OH. It is the simplest alcohol, and is a light, volatile, colorless, flammable, poisonous liquid with a distinctive odor that is somewhat milder and sweeter than ethanol (ethyl alcohol). At room temperature it is a polar solvent and is used as an antifreeze, solvent, fuel, and as a denaturant for ethyl alcohol. It is not popular for machinery, but may be found in automotive windshield washer fluid, de-icers, and gasoline additives.

Ethylene Glycol

Ethylene glycol

Ethylene glycol solutions became available in 1926 and were marketed as "permanent antifreeze" since the higher boiling points provided advantages for summertime use as well as during cold weather. They are used today for a variety of applications, including automobiles.

Poisoning

Ethylene glycol is toxic to humans, as well as to other animals, and should therefore be handled and disposed of properly. It has a sweet taste that can contribute to its accidental ingestion or its deliberate use as a murder weapon, as attributed by sensational media reports concerning it. Such poisoning is difficult to identify without specialized testing, as it often mimics other illnesses, and various symptoms can result from such poisoning, including severe diarrhea and vomiting. Some ethylene glycol antifreeze contains an embittering agent such as denatonium to help discourage either accidental or deliberate consumption.

Glycerol

Once used for automotive antifreeze, glycerol has the advantage of being non-toxic and could be used in the future for some applications.

Propylene Glycol

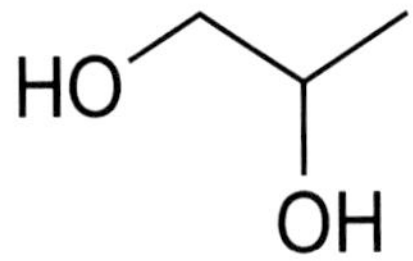

Propylene glycol

Propylene glycol, on the other hand, is considerably less toxic and may be labeled as "non-toxic antifreeze". It is used as antifreeze where ethylene glycol would be inappropriate, such as in food-processing systems or in water pipes in homes where incidental ingestion may be possible.

Propylene glycol oxidizes when exposed to air and heat, forming lactic acid. If not properly inhibited, this fluid can be very corrosive, so pH buffering agents are often added to propylene glycol, to prevent acidic corrosion of metal components. Besides cooling system breakdown, biological fouling also occurs. Once bacterial slime starts, the corrosion rate of the system increases. Maintenance of systems using glycol solution includes regular monitoring of freeze protection, pH, specific gravity, inhibitor level, colour, and biological contamination. Propylene glycol should be replaced when it turns a reddish colour.

Measuring the Freeze Point

Once antifreeze has been mixed with water and put into use, it periodically needs to be maintained. If engine coolant leaks, boils, or if the radiator needs to be drained and refilled, the antifreeze's freeze protection will need to be considered. In other cases a vehicle may need to be operated in a colder environment, requiring more antifreeze and less water. Three methods are commonly employed to determine the freeze point of the solution:

1. Specific gravity— (using a hydrometer or some sort of floating indicator),
2. Refractometer— which uses the optical properties of the solution, and
3. Test strips— specialized, disposable indicators made for this purpose.

Although ethylene glycol hydrometers are widely available and mass-marketed for antifreeze testing, they give false readings at high temperatures because specific gravity changes with temperature. Propylene glycol solutions cannot be tested using specific gravity because of ambiguous results (40% and 100% solutions have the same specific gravity).

Corrosion Inhibitors

Most commercial antifreeze formulations include corrosion inhibiting compounds, and a coloured dye (commonly a green, red, orange, yellow, or blue fluorescent) to aid in identification. A 1:1 dilution with water is usually used, resulting in a freezing point of about “34 °F (“37 °C), depending on the formulation. In warmer or colder areas, weaker or stronger dilutions are used, respectively, but a range of 40%/60% to 60%/40% is frequently specified to ensure corrosion protection, and 70%/30% for maximum freeze prevention down to “84 °F (“64 °C).

Traditional Inhibitors

Traditionally, there were two major corrosion inhibitors used in vehicles: silicates and phosphates. American made vehicles traditionally used both silicates and phosphates. European makes contain phosphates and other inhibitors, but no silicates. Japanese makes traditionally use phosphates and other inhibitors, but no silicates.

Organic Acid Technology

Certain cars are built with organic acid technology (OAT) antifreeze (e.g., DEX-COOL[16]), or with a hybrid organic acid technology (HOAT) formulation (e.g., Zerex G-05[16]), both of which are claimed to have an extended service life of five years or 240,000 km (150,000 mi).

DEX-COOL specifically has caused controversy. Litigation has linked it with intake manifold gasket failures in General Motors' (GM's) 3.1L and 3.4L engines, and with other failures in 3.8L and 4.3L engines. Class action lawsuits were registered in several states, and in Canada, to address some of these claims. The first of these to reach a decision was in Missouri where a settlement was announced early in December 2007. Late in March 2008, GM agreed to compensate complainants in the remaining 49 states. GM (Motors Liquidation Company) filed for bankruptcy in 2009, which tied up the outstanding claims until a court determines who gets paid.

According to the DEX-COOL manufacturer, “mixing a ‘green’ [non-OAT] coolant with DEX-COOL reduces the batch's change interval to 2 years or 30,000 miles, but will otherwise cause no damage to the engine.” DEX-COOL antifreeze uses one of two inhibitors: sebacate and 2-EHA (2-ethylhexanoic acid), which works well with the hard water found in the US, but is a plasticizer which can cause gaskets to leak. According to internal GM documents, the ultimate culprit appears to be operating vehicles for long periods of time with low coolant levels. The low coolant is caused by pressure caps that fail in the open position. (The new caps and recovery bottles were introduced at the same time as DEX-COOL). This exposes hot engine components to air and vapours, causing corrosion and contamination of the coolant with iron oxide particles, which in turn can aggravate the pressure cap problem as contamination holds the caps open permanently.

Honda and Toyota's new extended life coolant use OAT with sebacate as the primary OAT inhibitor, with some phosphates to provide protection while the OAT builds up. Honda specifically excludes 2-EHA from their formulas. Typically OAT antifreeze contains an orange dye to differentiate it from the conventional glycol-based coolants (green or yellow). Some of the newer OAT coolants claim to be compatible with *all* types of OAT and glycol-based coolants; these are typically green or yellow in colour.

Hybrid Organic Acid Technology

HOAT coolants typically mix an OAT with a traditional inhibitor, such as silicates or phosphates. G05 is a low-silicate, phosphate free formula that includes the benzoate inhibitor.

Additives

All automotive antifreeze formulations, including the newer organic acid (OAT antifreeze) formulations, are environmentally hazardous because of the blend of additives (around 5%), including lubricants, buffers and corrosion inhibitors. Because the additives in antifreeze are proprietary, the material safety data sheets (MSDS) provided by the manufacturer list only those compounds which are considered to be significant safety hazards when used in accordance with the manufacturer's recommendations. Common additives include sodium silicate, disodium phosphate, sodium molybdate, sodium borate, and dextrin (hydroxyethyl starch). Disodium fluorescein dyes are added to antifreeze to help trace the source of leaks, and as an identifier since some different formulations are incompatible.

Automotive antifreeze has a characteristic odor due to the additive tolytriazole, a corrosion inhibitor. The unpleasant odor in industrial use tolytriazole comes from impurities in the product that are formed from the toluidine isomers (ortho-, meta- and para-toluidine) and meta-diamino toluene which are side-products in the manufacture of tolytriazole. These side-products are highly reactive and produce volatile aromatic amines which are responsible for the unpleasant odor.

Polyester

Polyester is a category of polymers which contain the ester functional group in their main chain. Although there are many polyesters, the term "polyester" as a specific material most commonly refers to polyethylene terephthalate (PET). Polyesters include naturally occurring chemicals, such as in the cutin of plant cuticles, as well as synthetics through step-growth polymerization such as polycarbonate and polybutyrate. Natural polyesters and a few synthetic ones are biodegradable, but most synthetic polyesters are not.

Depending on the chemical structure, polyester can be a thermoplastic or thermoset; however, the most common polyesters are thermoplastics.

Fabrics woven or knitted from polyester thread or yarn are used extensively in apparel and home furnishings, from shirts and pants to jackets and hats, bed sheets, blankets, upholstered furniture and computer mouse mats. Industrial polyester fibres, yarns and ropes are used in tyre reinforcements, fabrics for conveyor belts, safety belts, coated fabrics and plastic reinforcements with high-energy absorption. Polyester fiber is used as cushioning and insulating material in pillows, comforters and upholstery padding.

While synthetic clothing in general is perceived by many as having a less natural feel compared to fabrics woven from natural fibres (such as cotton and wool), polyester fabrics can provide specific advantages over natural fabrics, such as improved wrinkle resistance, durability and high colour retention. As a result, polyester fibres are sometimes spun together with natural fibres to produce a cloth with blended properties. Synthetic fibres also can create materials with superior water, wind and environmental resistance compared to plant-derived fibres. Polyesters are also used to make "plastic" bottles, films, tarpaulin, canoes, liquid crystal displays, holograms, filters, dielectric film for capacitors, film insulation for wire and insulating tapes.

Liquid crystalline polyesters are among the first industrially used liquid crystal polymers. They are used for their mechanical properties and heat-resistance. These traits are also important in their application as an abradable seal in jet engines.

Polyesters are widely used as a finish on high-quality wood products such as guitars, pianos and vehicle/yacht interiors. Burns Guitars, Saab and Sunseeker are a few companies that use polyesters to finish their products. Thixotropic properties of spray-applicable polyesters make them ideal for use on open-grain timbers, as they can quickly fill wood grain, with a high-build film thickness per coat. Cured polyesters can be sanded and polished to a high-gloss, durable finish.

Types

Polyesters as thermoplastics may change shape after the application of heat. While combustible at high temperatures, polyesters tend to shrink away from flames and self-extinguish upon ignition. Polyester fibres have high tenacity and E-modulus as well as low water absorption and minimal shrinkage in comparison with other industrial fibres. Unsaturated polyesters (UPR) are thermosetting resins. They are used as casting materials, fiberglass laminating resins and non-metallic auto-body fillers. Fibreglass-reinforced unsaturated polyesters find wide application in bodies of yachts and as body parts of cars. According to the composition of their main chain, polyesters can be:

Composition of the main chain	*Number of repeating units*	*Examples of polyesters*	*Examples of manufacturing methods*
Aliphatic	Homopolymer	Polyglycolide or Polyglycolic acid (PGA)	Polycondensation of glycolic acid
		Polylactic acid (PLA)	Ring-opening polymerization of lactide
		Polycaprolactone (PCL)	Ring-opening polymerization of caprolactone
	Copolymer	Polyethylene adipate (PEA)	
		Polyhydroxyalkanoate (PHA)	
Semi-aromatic	Copolymer	Polyethylene terephtalate (PET)	Polycondensation of terephthalic acid with ethylene glycol
		Polybutylene terephthalate (PBT)	Polycondensation of terephthalic acid with 1,4-butanediol
		Polytrimethylene terephthalate (PTT)	Polycondensation of terephthalic acid with 1,3-propanediol
		Polyethylene naphthalate (PEN)	Polycondensation of at least one naphthalene dicarboxylic acid with ethylene glycol
Aromatic	Copolymer	Vectran	Polycondensation of 4-hydroxybenzoic acid and 6-hydroxynaphthalene-2-carboxylic acid

Increasing the aromatic parts of polyesters increases their glass transition temperature, melting temperature, thermal stability, chemical stability...

Polyesters can also be telechelic oligomers like the polycaprolactone diol (PCL) and the polyethylene adipate diol (PEA). They are then used as prepolymers.

Industry

Basics

Polyester is a synthetic polymer made of purified terephthalic acid (PTA) or its dimethyl ester dimethyl terephthalate (DMT) and monoethylene glycol (MEG). With 18% market share of all plastic materials produced, it ranges third after polyethylene (33.5%) and polypropylene (19.5%).

The main raw materials are described as follows:

- Purified terephthalic acid – PTA – CAS-No.: 100-21-0
  Synonym: 1,4 benzenedicarboxylic acid,
  Sum formula; $C_6H_4(COOH)_2$, mol weight: 166.13
- Dimethylterephthalate – DMT – CAS-No: 120-61-6
  Synonym: 1,4 benzenedicarboxylic acid dimethyl ester
  Sum formula $C_6H_4(COOCH_3)_2$, mol weight: 194.19
- Mono Ethylene Glycol – MEG – CAS No.: 107-21-1
  Synonym: 1,2 ethanediol
  Sum formula: $C_2H_6O_2$, mol weight: 62,07.

To make a polymer of high molecular weight a catalyst is needed. The most common catalyst is antimony trioxide (or antimony tri acetate):

Antimony trioxide – ATO – CAS-No.: 1309-64-4 Molecular weight: 291.51 Sum formula: Sb_2O_3

In 2008, about 10,000 tonnes Sb_2O_3 were used to produce around 49 million tonnes polyethylene terephthalate.

Polyester is described as follows:

Polyethylene Terephthalate CAS-No.: 25038-59-9 Synonym/ abbreviations: polyester, PET, PES Sum Formula: H-[$C_{10}H_8O_4$]-n=60–120 OH, molelcular unit weight: 192.17

There are several reasons for the importance of Polyester:

- The relatively easy accessible raw materials PTA or DMT and MEG
- The very well understood and described simple chemical process of polyester synthesis
- The low toxicity level of all raw materials and side products during polyester production and processing
- The possibility to produce PET in a closed loop at low emissions to the environment
- The outstanding mechanical and chemical properties of polyester
- The recyclability
- The wide variety of intermediate and final products made of polyester.

In table below the estimated world polyester production is shown. Main applications are textile polyester, bottle polyester resin, film polyester mainly for packaging and specialty polyesters for engineering plastics.

According to this table, the world's total polyester production might exceed 50 million tons per annum before the year 2010.

Table 5 : World polyester production

	Market size per year	
Product type	***2002 [Million tonnes/year]***	***2008 [Million tonnes/year]***
Textile-PET	20	39
Resin, bottle/A-PET	9	16
Film-PET	1.2	1.5
Special polyester	1	2.5
Total	31.2	49

Raw Material Producer

The raw materials PTA, DMT, and MEG are mainly produced by large chemical companies which are sometimes integrated down to the crude oil refinery where *p*-Xylene is the base material to produce PTA and liquefied petroleum gas (LPG) is the base material to produce MEG.

Polyester Processing

After the first stage of polymer production in the melt phase, the product stream divides into two different application areas which are mainly textile applications and packaging applications.

Table 6 : Textile and packaging polyester application list

Polyester-based polymer (melt or pellet)

Textile	***Packaging***
Staple fiber (PSF)	Bottles for CSD, Water, Beer, Juice, Detergents
Filaments POY, DTY, FDY	A-PET Film
Technical yarn and tire cord	Thermoforming
Non-woven and spunbond	BO-PET Biaxial oriented Film
Mono-filament	Strapping

Abbreviations: PSF = Polyester Staple Fiber; POY = Partially Oriented Yarn; DTY = Draw Textured Yarn; FDY = Fully Drawn Yarn; CSD = Carbonated Soft Drink; A-PET = Amorphous Polyester Film; BO-PET = Biaxial Oriented Polyester Film;

A comparable small market segment (much less than 1 million tonnes/year) of polyester is used to produce engineering plastics and masterbatch.

In order to produce the polyester melt with a high efficiency, high-output processing steps like staple fibre (50–300 tonnes/day per spinning line) or POY/FDY (up to 600 tonnes/day split into about 10 spinning machines) are meanwhile more and more horizontal, integrated, direct processes. This means the polymer melt is directly converted into the textile fibres or filaments without the common step of pelletizing. We are talking about full horizontal integration when polyester is produced at one site starting from crude oil or distillation products in the chain oil → benzene → PX → PTA → PET melt → fibre/filament or bottle-grade resin. Such integrated processes are meanwhile established in more or less interrupted processes at one production site. Eastman Chemicals introduced at first the idea to close the chain from PX to PET resin with their so-called INTEGREX process. The capacity of such horizontal, integrated productions sites is >1000 tonnes/day and can easily reach 2500 tonnes/day.

Besides the above mentioned large processing units to produce staple fibre or yarns, there are ten thousands of small and very small processing plants, so that one can estimate that polyester is processed and recycled in more than 10 000 plants around the globe. This is without counting all the companies involved in the supply industry, beginning with engineering and processing machines and ending with special additives, stabilizers and colours. This is a gigantic industry complex and it is still growing by 4–8% per annum, depending on the world region. Useful information about the polyester industry can be found under where a "Who is Producing What in the Polyester Industry" is gradually being developed.

Synthesis

Synthesis of polyesters is generally achieved by a polycondensation reaction. The general equation for the reaction of a diol with a diacid is :

$$(n+1)\ R(OH)_2 + n\ R'(COOH)_2 \rightarrow HO[ROOCR'COO]_nROH + 2n\ H_2O$$

Azeotrope Esterification

In this classical method, an alcohol and a carboxylic acid react to form a carboxylic ester. To assemble a polymer, the water formed by the reaction must be continually removed by azeotrope distillation.

Alcoholic Transesterification

Transesterification: An alcohol-terminated oligomer and an ester-terminated oligomer condense to form an ester linkage, with loss of an alcohol. R and R' are the two oligomer chains, R" is a sacrificial unit such as a methyl group (methanol is the byproduct of the esterification reaction).

Acylation (HCl Method)

The acid begins as an acid chloride, and thus the polycondensation proceeds with emission of hydrochloric acid (HCl) instead of water. This method can be carried out in solution or as an enamel.

Silyl Method

In this variant of the HCl method, the carboxylic acid chloride is converted with the trimethyl silyl ether of the alcohol component and production of trimethyl silyl chloride is obtained

Ring-opening Polymerization

Aliphatic polyesters can be assembled from lactones under very mild conditions, catalysed anionically, cationically or metallorganically.

Cross-linking

Unsaturated polyesters are thermosetting resins. They are generally copolymers prepared by polymerizing one or more diol with saturated and unsaturated dicarboxylic acids (maleic acid, fumaric acid...) or their anhydrides. The double bond of unsaturated polyesters reacts with a vinyl monomer mainly the styrene, resulting in a 3-D cross-linked structure. This structure acts as a thermoset. The cross-linking is initiated through an exothermic reaction involving an organic peroxide, such as methyl ethyl ketone peroxide or benzoyl peroxide.

Glycol Ethers

Glycol ethers are a group of solvents based on alkyl ethers of ethylene glycol commonly used in paints. These solvents typically have higher boiling point, together with the favourable solvent properties of lower molecular weight of ethers and alcohols. The word “Cellosolve” was registered in 1924 as a United States trademark by Carbide & Carbon Chemicals Corp. (later named Union Carbide Corp.) for “Solvents for Gums, Resins, Cellulose Esters, and the Like”, ; the first one was ethyl cellosolve (ethylene glycol monoethyl ether), with the name now generic for glycol ethers.

Glycol ethers can be also derived of diethylene glycol (carbitols). Acetates of glycols are a similar kind of potent solvents. Recent study suggests that occupational exposure to glycol ethers is related to low motile sperm count in men, but the finding has been disputed by others.

Glycol Ether Solvents

- Ethylene glycol monomethyl ether (2-methoxyethanol, $CH_3OCH_2CH_2OH$)
- Ethylene glycol monoethyl ether (2-ethoxyethanol, $CH_3CH_2OCH_2CH_2OH$)
- Ethylene glycol monopropyl ether (2-propoxyethanol, $CH_3CH_2CH_2OCH_2CH_2OH$)

- Ethylene glycol monoisopropyl ether (2-isopropoxyethanol, $(CH_3)_2CHOCH_2CH_2OH$)
- Ethylene glycol monobutyl ether (2-butoxyethanol, $CH_3CH_2CH_2CH_2OCH_2CH_2OH$), a widely used solvent in paintings and surface coatings, cleaning products and inks
- Ethylene glycol monophenyl ether (2-phenoxyethanol, $C_6H_5OCH_2CH_2OH$)
- Ethylene glycol monobenzyl ether (2-benzyloxyethanol, $C_6H_5CH_2OCH_2CH_2OH$)
- Diethylene glycol monomethyl ether (2-(2-methoxyethoxy) ethanol, methyl carbitol, $CH_3OCH_2CH_2OCH_2CH_2OH$)
- Diethylene glycol monoethyl ether (2-(2-ethoxyethoxy)ethanol, carbitol cellosolve, $CH_3CH_2OCH_2CH_2OCH_2CH_2OH$)
- Diethylene glycol mono-n-butyl ether (2-(2-butoxyethoxy) ethanol, $CH_3CH_2CH_2CH_2OCH_2CH_2OCH_2CH_2OH$)

Dialkyl Ethers

- Ethylene glycol dimethyl ether (dimethoxyethane, $CH_3OCH_2CH_2OCH_3$), a higher boiling alternative to diethyl ether and THF, also used as a solvent for polysaccharides, a reagent in organometallic chemistry and in some electrolytes of lithium batteries
- Ethylene glycol diethyl ether (diethoxyethane, $CH_3CH_2OCH_2CH_2OCH_2CH_3$)
- Ethylene glycol dibutyl ether (dibutoxyethane, $CH_3CH_2CH_2CH_2OCH_2CH_2OCH_2CH_2CH_2CH_3$)

Esters

- Ethylene glycol methyl ether acetate (2-methoxyethyl acetate, $CH_3OCH_2CH_2OCOCH_3$)
- Ethylene glycol monethyl ether acetate (2-ethoxyethyl acetate, $CH_3CH_2OCH_2CH_2OCOCH_3$)
- Ethylene glycol monobutyl ether acetate (2-butoxyethyl acetate, $CH_3CH_2CH_2CH_2OCH_2CH_2OCOCH_3$)

4

Narrow Range Ethoxylate

Fatty alcohol polyglycol ethers with a narrow homolog distribution, which are also known as narrow-range ethoxylates (NRE), are known nonionic surfactants which are industrially produced, for example, by addition of ethylene oxide onto fatty alcohols in the presence of suitable catalysts (layer compounds which have been calcined or hydrophobized with fatty acids). This process can also be carried out on a variety of other hydrophobes and using different alkoxylating compounds (e.g., propylene oxide and butylene oxide) by modifying the catalyst properties.

For example:

An ethoxylation reaction proceeds under an inert atmosphere with an amount of heat depending on the starting material. The reaction proceeds via the epoxide (in this case ethylene oxide) ring opening and activation of the nucleophile, ring, or combination thereof via the catalyst.

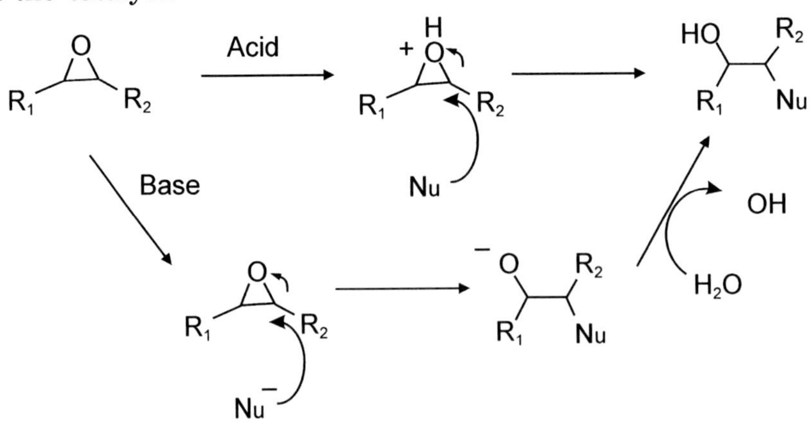

With conventional catalysts (i.e. potassium hydroxide, sodium hydroxide, etc.) one obtains a distribution of ethoxymers. If one targets a 12 mole ethoxylate with a conventional alkaline earth hydroxide, one could expect to get a broad range of ethoxylates with n being anywhere from 3 to 30 moles of EO. With narrow range catalysts a much tighter distribution can be obtained.

Narrow range surfactants like alcohol ethoxylates and propoxylates can be incorporated into alkyl ether sulfates or mixed with other anionic surfactants and exhibit beneficial properties such as reduced irritation to skin/eyes and lower free alcohol content.

Vinyl Acetate

Vinyl acetate is an organic compound with the formula $CH_3COOCH=CH_2$. A colorless liquid with a pungent odor, it is the precursor to polyvinyl acetate, an important polymer in industry.

Production

The worldwide production capacity of vinyl acetate monomer (VAM) was estimated at 6,154,000 tonnes/annum in 2007, with most capacity concentrated in the United States (1,585,000 all in Texas), China (1,261,000), Japan (725,000) and Taiwan (650,000). The average list price for 2008 was $1600/tonne. Celanese is the largest producer (ca 25% of the worldwide capacity), while other significant producers include China Petrochemical Corporation (7%), Chang Chun Group (6%) and LyondellBasell (5%).

Preparation

The major industrial route involves the reaction of ethylene and acetic acid with oxygen in the presence of a palladium catalyst.

Ethylene + acetic acid + 1/2 $O_2 \rightarrow$ Vinyl acetate + H_2O

But byproducts are also generated:

Ethylene + 3 $O_2 \rightarrow 2\ CO_2 + 2\ H_2O$

Vinyl acetate is also prepared by the gas-phase addition of acetic acid to acetylene.

Polymerization

It can be polymerized, either by itself to make polyvinyl acetate or with other monomers to prepare copolymers such as ethylene-vinyl acetate (EVA), vinyl acetate-acrylic acid (VA/AA) and polyvinyl chloride acetate (PVCA). Due to the instability of the radical, attempts to control the polymerization via most 'living/controlled' radical processes have proved problematic. However, RAFT (or more specifically MADIX) polymerization offers a convenient method of controlling the synthesis of PVA by the addition of a xanthate chain transfer agent.

Other Reactions

Vinyl acetate undergoes many of the reactions anticipated for an alkene and an ester. Bromine adds to give the dibromide. Hydrogen halides add to give 1-haloethyl acetates, which cannot be generate by other methods because of the non-availability of the corresponding halo-alcohols. Acetic acid adds in the presence of palladium catalysts to give ethylidene diacetate, $CH_3CH(OAc)_2$. It undergoes transesterification with a variety of carboxylic acids. The alkene also undergoes Diels-Alder and 2+2 cycloadditions.

Toxicity Evalution

On January 31, 2009, the Government of Canada's final assessment concluded that exposure to vinyl acetate is not considered to be harmful to human health. This decision under the Canadian Environmental Protection Act (CEPA) was based on new information received during the public comment period, as well as more recent information from the risk assessment conducted by the European Union.

1,2-Dichloroethane

The chemical compound 1,2-dichloroethane, commonly known by its old name of ethylene dichloride (EDC), is a chlorinated hydrocarbon, mainly used to produce vinyl chloride monomer (VCM, chloroethene), the major precursor for PVC production. It is a colourless liquid with a chloroform-like odour. 1,2-Dichloroethane is also used generally as an intermediate for other organic chemical compounds and as a solvent. It forms azeotropes with many other solvents, including water (b.p. 70.5 C) and other chlorocarbons.

History

In 1794, physician Jan Rudolph Deiman, merchant Adriaan Paets van Troostwijk, chemist Anthoni Lauwerenburg, and botanist Nicolaas Bondt, under the name of *Gezelschap der Hollandsche Scheikundigen* (Dutch: *Society of Dutch Chemists*), were the first to produce 1,2-

dichloroethane from olefiant gas (oil-making gas, ethylene) and chlorine gas. Although the *Gezelschap* in practice did not do much in-depth scientific research, they and their publications were highly regarded. Part of that acknowledgement is that 1,2-dichloroethane has been called "Dutch oil" in old chemistry.

Production

Nearly 20 million tons of 1,2-dichloroethane are produced in the United States, Western Europe, and Japan. Production is primarily achieved through the iron(III) chloride-catalysed reaction of ethene (ethylene) and chlorine.

$$H_2C{=}CH_2 + Cl_2 \rightarrow ClCH_2\text{-}CH_2Cl$$

1,2-dichloroethane is also generated by the copper(II) chloride-catalysed "oxychlorination" of ethylene:

$$2\ H_2C{=}CH_2 + 4\ HCl + O_2 \rightarrow 2\ ClCH_2\text{-}CH_2Cl + 2\ H_2O$$

In principle, it can be prepared by the chlorination of ethane and, less directly, from ethanol.

Uses

Vinyl Chloride Monomer (VCM) Production

With approximately 80% of the world's consumption of 1,2-dichloroethane, the major use of 1,2-dichloroethane is in the production of vinyl chloride monomer (VCM, chloroethene) with hydrogen chloride as a byproduct. VCM is the precursor to polyvinyl chloride.

$$Cl\text{-}CH_2\text{-}CH_2\text{-}Cl \rightarrow H_2C{=}CH\text{-}Cl + HCl$$

The hydrogen chloride can be re-used in the production of more 1,2-dichloroethane via the oxychlorination route described above.

Other Uses

As a good apolar aprotic solvent, 1,2-dichloroethane is used as degreaser and paint remover. As a useful 'building block' reagent, it is used as an intermediate in the production of various organic compounds such as ethylenediamine. In the laboratory it is occasionally used as a source of chlorine, with elimination of ethene and chloride.

Via several steps, 1,2-dichloroethane is a precursor to 1,1,1-trichloroethane, which is used in dry cleaning. Historically, 1,2-dichloroethane was used as an anti-knock additive in leaded fuels.

Safety

1,2-Dichloroethane is toxic (especially by inhalation due to its high vapour pressure), corrosive, highly flammable, and carcinogenic. Its high solubility and 50-year half-life in anoxic aquifers make it a perennial pollutant and health risk that is very expensive to treat conventionally, requiring a method of bioremediation. While the chemical is banned from use by U.S. manufacturers, a case was reported in 2009 of molded plastic consumer products (toys and holiday decorations) from China that released 1,2-dichloroethane into homes at levels high enough to produce cancer risk. Substitutes are recommended and will vary according to application. Dioxolane and toluene are possible substitutes as solvents. Dichloroethane is unstable in the presence of aluminium metal and, when moist, with zinc and iron.

Trichloroethylene

The chemical compound trichloroethylene (C_2HCl_3) is a chlorinated hydrocarbon commonly used as an industrial solvent. It is a clear non-flammable liquid with a sweet smell.

The IUPAC name is trichloroethene. Industrial abbreviations include TCE, trichlor, Trike, Tricky and tri. It has been sold under a variety of trade names. Under the trade names Trimar and Trilene, trichloroethylene was used as a volatile anesthetic and as an inhaled obstetrical analgesic in millions of patients.

History

Pioneered by Imperial Chemical Industries in Britain, its development was hailed as an anesthetic revolution. Originally thought to possess less hepatotoxicity than chloroform, and without the unpleasant pungency and flammability of ether, TCE use was nonetheless soon found to have several pitfalls. These included promotion of cardiac arrhythmias, too low a volatility for quick anesthetic induction, reactions with soda lime used in carbon dioxide absorbing systems, prolonged neurologic dysfunction when used with soda lime, and evidence of hepatotoxicity as had been found with chloroform.

The introduction of halothane in 1956 greatly diminished the use of TCE as a general anesthetic. TCE was still used as an inhalation analgesic in childbirth given by self-administration. Fetal toxicity and concerns for carcinogenic potential of TCE led to its abandonment in the 1980s.

Due to concerns about its toxicity, the use of trichloroethylene in the food and pharmaceutical industries has been banned in much of the world since the 1970s. Legislation has forced the substitution of trichloroethylene in many processes in Europe as the chemical was classified as a carcinogen carrying an R45 risk phrase. Many degreasing chemical alternatives are being promoted such as Ensolv and Leksol; however, each of these is based on *n*-propyl bromide which carries an R60 risk phrase and they would not be a legally acceptable substitute.

Groundwater contamination by TCE has become an important environmental concern.

Recently it was announced by the USEPA (the United States Environmental Protection Agency) that the agency had completed its Final Health Assessment for Trichloroethylene and released a list of new TCE toxicity values. The results of the study have formally characterized the chemical as a human carcinogen and a non-carcinogenic health hazard.

Production

Prior to the early 1970s, most trichloroethylene was produced in a two-step process from acetylene. First, acetylene was treated with chlorine using a ferric chloride catalyst at 90 °C to produce 1,1,2,2-tetrachloroethane according to the chemical equation:

$$HCa^{-}CH + 2\ Cl_2 \rightarrow Cl_2CHCHCl_2$$

The 1,1,2,2-tetrachloroethane is then dehydrochlorinated to give trichloroethylene. This can either be accomplished with an aqueous solution of calcium hydroxide

$$2\ Cl_2CHCHCl_2 + Ca(OH)_2 \rightarrow 2\ ClCH{=}CCl_2 + CaCl_2 + 2\ H_2O$$

or in the vapour phase by heating it to 300-500°C on a barium chloride or calcium chloride catalyst

$$Cl_2CHCHCl_2 \rightarrow ClCH{=}CCl_2 + HCl$$

Today, however, most trichloroethylene is produced from ethylene. First, ethylene is chlorinated over a ferric chloride catalyst to produce 1,2-dichloroethane.

$$CH_2{=}CH_2 + Cl_2 \rightarrow ClCH_2CH_2Cl$$

When heated to around 400 °C with additional chlorine, 1,2-dichloroethane is converted to trichloroethylene

$$ClCH_2CH_2Cl + 2\ Cl_2 \rightarrow ClCH{=}CCl_2 + 3\ HCl$$

This reaction can be catalysed by a variety of substances. The most commonly used catalyst is a mixture of potassium chloride and aluminum chloride. However, various forms of porous carbon can also be used. This reaction produces tetrachloroethylene as a byproduct, and depending on the amount of chlorine fed to the reaction, tetrachloroethylene can even be the major product. Typically, trichloroethylene and tetrachloroethylene are collected together and then separated by distillation.

Uses

Trichloroethylene is an effective solvent for a variety of organic materials.

When it was first widely produced in the 1920s, trichloroethylene's major use was to extract vegetable oils from plant materials such as soy, coconut, and palm. Other uses in the food industry included coffee decaffeination and the preparation of flavouring extracts from hops and spices. It has also been used for drying out the last bit of water for production of 100% ethanol.

From the 1930s through the 1970s, both in Europe and North America, trichloroethylene was used as a volatile gas anesthetic. TCE was used in place of the earlier anesthetics chloroform and ether in the 1940s, but was itself replaced in the 1950s by the newer halothane, which allowed much faster induction and recovery times. Marketed in the UK by ICI under the trade name Trilene it was coloured blue (with a dye called waxolene blue) to avoid confusion with the similar smelling chloroform.

It has also been used as a dry cleaning solvent, although replaced in the 1950s by tetrachloroethylene (also known as perchloroethylene).

Trichloroethylene was marketed as 'Ecco 1500 Anti-Static Film Cleaner and Conditioner' until 2009, for use in automatic movie film cleaning machines, and for manual cleaning with lint-free wipes.

Perhaps the greatest use of TCE has been as a degreaser for metal parts. The demand for TCE as a degreaser began to decline in the 1950s in favour of the less toxic 1,1,1-trichloroethane. However, 1,1,1-trichloroethane production has been phased out in most of the world under the terms of the Montreal Protocol, and as a result trichloroethylene has experienced some resurgence in use as a degreaser.

TCE has also been used in the United States to clean kerosene-fuelled rocket engines (TCE was not used to clean hydrogen-fuelled engines such as the Space Shuttle Main Engine). During static firing, the RP-1 fuel would leave hydrocarbon deposits and vapours in the engine. These deposits had to be flushed from the engine to avoid the possibility of explosion during engine handling and future firing. TCE was used to flush the engine's fuel system immediately before and after each test firing. The flushing procedure involved pumping TCE through the engine's fuel system and letting the solvent overflow for a period ranging from several seconds to 30–35 minutes, depending upon the engine. For some engines, the engine's gas generator and LOX dome were also flushed with TCE prior to test firing. The F-1 rocket engine had its LOX dome, gas generator, and thrust chamber fuel jacket flushed with TCE during launch preparations.

TCE is also used in the manufacture of a range of fluorocarbon refrigerants such as 1,1,1,2-Tetrafluoroethane more commonly known as HFC 134a

Chemical Instability

Despite its widespread use as a metal degreaser, trichloroethylene itself is unstable in the presence of metal over prolonged exposure. As early as 1961 this phenomenon was recognized by the manufacturing industry, when stabilizing additives were added in the commercial formulation. Since the reactive instability is accentuated by higher temperatures, the search for stabilizing additives was conducted by heating trichloroethylene to its boiling point in a reflux condenser and observing decomposition. The first widely used stabilizing additive was dioxane; however, its use was patented by Dow Chemical Company and could not be used by other manufacturers. Considerable research took place in the 1960s to develop alternative stabilizers for trichloroethylene. Other chemical stabilizers include ketones such as methyl ethyl ketone.

Physiological Effects

When inhaled, trichloroethylene produces central nervous system depression resulting in general anesthesia. Its high blood solubility results in a less desirable slower induction of anesthesia. At low concentrations it is relatively non-irritating to the respiratory tract. Higher concentrations result in tachypnea. Many types of cardiac arrhythmias can occur and are exacerbated by epinephrine (adrenaline). It was noted in the 1940s that TCE reacted with carbon dioxide (CO_2) absorbing systems (soda lime) to produce dichloroacetylene and phosgene. Cranial nerve dysfunction (especially the fifth cranial nerve) was not uncommon when TCE anesthesia was given using CO_2 absorbing systems. These nerve deficits could last for months. Occasionally facial numbness was permanent. Muscle relaxation with TCE anesthesia sufficient for surgery was poor. For these reasons as well as problems with hepatotoxicity, TCE lost popularity in North America and Europe to more potent anesthestics such as halothane by the 1960s.

The symptoms of acute non-medical exposure are similar to those of alcohol intoxication, beginning with headache, dizziness, and confusion and progressing with increasing exposure to unconsciousness. Respiratory and circulatory depression can result in death.

Much of what is known about the human health effects of trichloroethylene is based on occupational exposures. Beyond the effects to the central nervous system, workplace exposure to trichloroethylene has been associated with toxic effects in the liver and kidney. Over time, occupational exposure limits on trichloroethylene have tightened, resulting in more stringent ventilation controls and personal protective equipment use by workers.

Research from Cancer bioassays performed by the National Cancer Institute (later the National Toxicology Program) showed that exposure to trichloroethylene is carcinogenic in animals, producing liver cancer in mice, and kidney cancer in rats. Research published in 1994 examined the incidence of leukemia and non-Hodgkin lymphoma in populations exposed to TCE in their drinking water.

The National Toxicology Program's 11th Report on Carcinogens categorizes trichloroethylene as "reasonably anticipated to be a human carcinogen", based on limited evidence of carcinogenicity from studies in humans and sufficient evidence of carcinogenicity from studies in experimental animals.

One recent review of the epidemiology of kidney cancer rated cigarette smoking and obesity as more important risk factors for kidney cancer than exposure to solvents such as trichloroethylene. In contrast, the most recent overall assessment of human health risks associated with trichloroethylene states, "[t]here is concordance between animal and human studies, which supports the conclusion that trichloroethylene is a potential kidney carcinogen". The evidence appears to be less certain at this time regarding the relationship between humans and liver cancer observed in mice, with the NAS suggesting that low-level exposure might not represent a significant liver cancer risk in the general population.

Recent studies in laboratory animals and observations in human populations suggest that exposure to trichloroethylene might be associated with congenital heart defects While it is not clear what levels of exposure are associated with cardiac defects in humans, there is consistency between the cardiac defects observed in studies of communities exposed to trichloroethylene contamination in groundwater, and the effects observed in laboratory animals. A study published in August 2008, has demonstrated effects of TCE on human mitochondria. The article questions whether this might impact female reproductive function.

The health risks of trichloroethylene have been studied extensively. The U.S. Environmental Protection Agency (EPA) sponsored a "state of the science" review of the health effects associated with exposure to trichloroethylene. The National Academy of Sciences concluded that evidence on the carcinogenic risk and other potential health hazards from exposure to TCE has strengthened since EPA released their toxicological assessment of TCE, and encourages federal agencies to finalize the risk assessment for TCE using currently available information, so that risk management decisions for this chemical can be expedited.

In Europe, the Scientific Committee on Occupational Exposure Limits (SCOEL) recommends for trichloroethylene an occupational exposure limit (8h time-weighted average) of 10 ppm and a short-term exposure limit (15 min) of 30 ppm.

Human Exposure

Some are exposed to TCE through contaminated drinking water. With a specific gravity greater than 1, trichloroethene can be present as a dense nonaqueous phase liquid if sufficient quantities are spilled in the environment. Another significant source of vapour exposure in Superfund sites that had contaminated groundwater, such as the

Twin Cities Army Ammunition Plant, was by showering. TCE readily volatilizes out of hot water and into the air. Long, hot showers would then volatilize more TCE into the air. In a home closed tightly to conserve the cost of heating and cooling, these vapours would then recirculate.

The first known report of TCE in groundwater was given in 1949 by two English public chemists who described two separate instances of well contamination by industrial releases of TCE. Based on available federal and state surveys, between 9% to 34% of the drinking water supply sources tested in the U.S. may have some TCE contamination, though EPA has reported that most water supplies are in compliance with the Maximum Contaminant Level (MCL) of 5 ppb. In addition, a growing concern in recent years at sites with TCE contamination in soil or groundwater has been vapour intrusion in buildings, which has resulted in indoor air exposures, such is in a recent case in the McCook Field Neighborhood of Dayton, Ohio. Trichloroethylene has been detected in 852 Superfund sites across the United States, according to the Agency for Toxic Substances and Disease Registry (ATSDR). Under the Safe Drinking Water Act of 1974, and as amended annual water quality testing is required for all public drinking water distributors. The EPA'S current guidelines for TCE can be found here. It should be noted that the EPA's table of "TCE Releases to Ground" is dated 1987 to 1993, thereby omitting one of the largest Superfund Cleanup sites in the nation, the NIBW in Scottsdale, Arizona. The TCE "released" here occurred prior to its appearance in the municipal drinking wells in 1982.

In 1998, the View-Master factory supply well in Beaverton, Oregon was found to have been contaminated with high levels of TCE. It was estimated that 25,000 factory workers had been exposed to it from 1950–2001.

As of 2007, 57,000 pounds, or roughly 19 tons of TCE have been removed from the system of wells that once supplied drinking water to the residents of Scottsdale. One of the three drinking water wells previously owned by the City of Phoenix and ultimately sold to the City of Scottsdale, tested at 390 ppb TCE when it was closed in 1982. Some Scottsdale residents who received their water bills from the City of Phoenix throughout the 1960s and 70's were understandably confused as to whether they indeed had been consuming contaminated water when information about the Superfund site was first disseminated.

The City of Scottsdale recently updated their website to clarify that the contaminated wells were "in the Scottsdale area" and to delete all references to the levels of TCE discovered when the wells were closed as "trace".

A spot was then ultimately chosen to receive and treat the contaminated drinking water known as the Central Groundwater Treatment Facility. Then 1989, as now, this treatment facility (CGTF) is situated on land adjacent to Pima Park and the Siemens facility documented as one of the Potentially Responsible Parties at the corner of Thomas and Pima roads. Close proximity to this park did not appear to enter into Motorola's calculations when asserting that it would save money to remove the carbon air filters in 2007.

Marine Corps Base Camp Lejeune in North Carolina may be the largest TCE contamination site in the country. Legislation could force the EPA to establish a health advisory and a national public drinking water regulation to limit trichloroethylene.

For over twenty years of operation, the US-based multinational Radio Company of America (RCA) had been pouring toxic wastewater into a well in its Taoyuan, Taiwan facility. The pollution from the plant was not revealed until 1994, when former workers brought it to light. Investigation by the Taiwan Environmental Protection Administration confirmed that RCA had been dumping chlorinated organic solvents into a secret well and caused contamination to the soil and groundwater surrounding the plant site. High levels of TCE tetrachloroethylene (PCE) can be found in groundwater drawn as far as two kilometres from the site. An organization of former RCA employees reports 1375 cancer cases, 216 cancer deaths, and 102 cases of various tumors among its members.

Trichloroethylene is a cleaning solvent that was used to clean military weapons during the Gulf War. There are reports associating exposure to this solvent with amyotrophic lateral sclerosis, and also with a neurologic syndrome resembling Parkinson's disease.

Existing Regulation

Until recent years, the US Agency for Toxic Substances and Disease Registry (ATSDR) contended that trichloroethylene had little-to-no carcinogenic potential, and was probably a co-carcinogen—that is, it acted in concert with other substances to promote the formation of tumours.

Half a dozen state, federal, and international agencies now classify trichloroethylene as a probable carcinogen. The International Agency for Research on Cancer considers trichloroethylene a Group 2A carcinogen, indicating that it considers it is probably carcinogenic to humans. California EPA regulators consider it a known carcinogen and issued a risk assessment in 1999 that concluded that it was far more toxic than previous scientific studies had shown.

Proposed U.S. Federal Regulation

In 2001, a draft report of the Environmental Protection Agency (EPA) laid the groundwork for tough new standards to limit public exposure to trichloroethylene. The assessment set off a fight between the EPA and the Department of Defence (DoD), the Department of Energy, and NASA, who appealed directly to the White House. They argued that the EPA had produced junk science, its assumptions were badly flawed, and that evidence exonerating the chemical was ignored.

The DoD has about 1,400 military properties nationwide that are contaminated with trichloroethylene. Twenty three sites in the Energy Department's nuclear weapons complex — including Lawrence Livermore National Laboratory in the San Francisco Bay area, and NASA centers, including the Jet Propulsion Laboratory in La Cañada Flintridge are reported to have TCE contamination.

Political appointees in the EPA sided with the Pentagon and agreed to pull back the risk assessment. In 2004, the National Academy of Sciences was given a $680,000 contract to study the matter, releasing its report in the summer of 2006. The report has raised more concerns about the health effects of TCE.

In response to the heightened awareness of environmental toxins such as TCE and the role they may be playing in childhood disease, Sen. Obama proposed S1068, cosponsored by Hillary Clinton and others. This legislation aims to inform and protect communities that are threatened with environmental contamination. Sen. Clinton's own bill, S1911, is known as the TCE Reduction Act. This bill was co-sponsored by Sen. Elizabeth Dole (R-North Carolina).

Reduced Production and Remediation

In recent times, there has been a substantial reduction in the production output of trichloroethylene; alternatives for use in metal degreasing abound, chlorinated aliphatic hydrocarbons being phased out in a large majority of industries due to the potential for irreversible health effects and the legal liability that ensues as a result.

The U.S. military has virtually eliminated its use of the chemical, purchasing only 11 gallons in 2005. About 100 tons of it is used annually in the U.S. as of 2006.

Recent research has focused on the in-place remediation of trichloroethylene in soil and ground water instead of removal for off-site treatment and disposal. Naturally-occurring bacteria have been identified with the ability to degrade or completely mineralize TCE. *Dehalococcoide* sp. degrade trichloroethylene by reductive dechlorination under anaerobic conditions. Under aerobic conditions, *Pseudomonas fluorescence* can co-metabolize TCE. Soil and ground water contamination by TCE has also been successfully remediated by chemical treatment and extraction.

Media

In the John Travolta film A Civil Action, the case of a TCE-contaminated tannery and town water supply is covered in detail.

Tetrachloroethylene

Tetrachloroethylene, also known under its systematic name tetrachloroethene and many other names, is a chlorocarbon with the formula $Cl_2C{=}CCl_2$. It is a colourless liquid widely used for dry cleaning of fabrics, hence it is sometimes called "dry-cleaning fluid." It has a sweet odor detectable by most people at a concentration of 1 part per million (1 ppm). Worldwide production was about 1 megatonne in 1985.

Production

Michael Faraday first synthesized tetrachloroethene in 1821 by thermal decomposition of hexachloroethane.

$$C_2Cl_6 \rightarrow C_2Cl_4 + Cl_2$$

Most tetrachloroethene is produced by high temperature chlorinolysis of light hydrocarbons. The method is related to Faraday's discovery since hexachloroethane is generated and thermally decomposes. Side products include carbon tetrachloride, hydrogen chloride, and hexachlorobutadiene.

Several other methods have been developed. When 1,2-dichloroethane is heated to 400 °C with chlorine, tetrachloroethene is produced by the chemical reaction:

$$ClCH_2CH_2Cl + 3\ Cl_2 \rightarrow Cl_2C{=}CCl_2 + 4\ HCl$$

This reaction can be catalysed by a mixture of potassium chloride and aluminium chloride or by activated carbon. Trichloroethylene is a major byproduct, which is separated by distillation.

According to an EPA report of 1976, the quantity of Tetrachloroethylene (also known as perchloroethylene or PCE) produced in the United States in just one year 1973, totaled 706 million pounds (320,000 metric tons). Diamond Shamrock, Dow Chemical Compnay, E.I DuPont and Vulcan Materials Company (Chemical Division) were among the top eight producers nationwide.

Uses

Tetrachloroethylene is an excellent solvent for organic materials. Otherwise it is volatile, highly stable, and nonflammable. For these reasons, it is widely used in dry cleaning. Usually as a mixture with other chlorocarbons, it is also used to degrease metal parts in the automotive and other metalworking industries. It appears in a few consumer products including paint strippers and spot removers.

Historical Applications

Tetrachloroethene was once extensively used as an intermediate in the manufacture of HFC-134a and related refrigerants. In the early 20th century, tetrachloroethene was used for the treatment for hookworm infestation.

Health and Safety

The International Agency for Research on Cancer has classified tetrachloroethene as a Group 2A carcinogen, which means that it is probably carcinogenic to humans. Like many chlorinated hydrocarbons, tetrachloroethene is a central nervous system depressant and can enter the body through respiratory or dermal exposure. Tetrachloroethene dissolves fats from the skin, potentially resulting in skin irritation.

Animal studies and a study of 99 twins by Dr. Samuel Goldman and researchers at the Parkinson's Institute in Sunnyvale, California determined there is a "lot of circumstantial evidence" that exposure to Tetrachloroethlene increases the risk of developing Parkinson's disease ninefold. Larger population studies are planned. At temperatures over 600 °F (316 °C), such as in welding, tetrachloroethylene can decompose into phosgene, an extremely poisonous gas. Tetrachloroethylene should not be used near welding operations, flames, or hot surfaces.

Testing for Exposure

Tetrachloroethene exposure can be evaluated by a breath test, analogous to breath-alcohol measurements. Because it is stored in the body's fat and slowly released into the bloodstream, tetrachloroethene can be detected in the breath for weeks following a heavy exposure. Tetrachloroethylene and trichloroacetic acid (TCA), a breakdown product of tetrachloroethene, can be detected in the blood.

In Europe, the Scientific Committee on Occupational Exposure Limits (SCOEL) recommends for tetrachloroethylene an occupational exposure limit (8h time-weighted average) of 20 ppm and a short-term exposure limit (15 min) of 40 ppm.

Environmental Contamination

Tetrachloroethene is a common soil contaminant. With a specific gravity greater than 1, tetrachloroethylene will be present as a dense nonaqueous phase liquid if sufficient quantities of liquid are spilled in the environment. Because of its mobility in groundwater, its toxicity at low levels, and its density (which causes it to sink below the water table), cleanup activities are more difficult than for oil spills. Recent research has focused on the in place remediation of soil and ground water pollution by tetrachloroethylene. Instead of excavation or extraction for above-ground treatment or disposal, tetrachloroethylene contamination has been successfully remediated by chemical treatment or bioremediation. Bioremediation has been successful under anaerobic conditions by reductive dechlorination by *Dehalococcoides* sp. and under aerobic conditions by cometabolism by *Pseudomonas* sp. Partial degradation daughter products include trichloroethylene, cis-1,2-dichloroethene and vinyl chloride; full degradation converts tetrachloroethylene to ethene and chloride dissolved in water.

It has been estimated that about 85% of tetrachloroethylene is released into the atmosphere; OECD models assumed 90% release into the air and 10% to water. Based on these models, its distribution in the environment is estimated to be in the air (76.39% - 99.69%), water (0.23% - 23.2%), soil (0.06-7%), with the remainder in the sediment and biota. Estimates of lifetime in the atmosphere vary, but a 1987 survey estimated the lifetime in the air has been estimated at about 2 months in the Southern Hemisphere and 5–6 months in the Northern Hemisphere. Degradation products observed in a laboratory include phosgene, trichloroacetyl chloride, hydrogen chloride, carbon dioxide, and carbon monoxide. In water, tetrachloroethylene is degraded very slowly by hydrolysis, and it is persistent under aerobic conditions. It is degraded through reductive dechlorination under anaerobic conditions, with the degradation products including trichloroethene, dichloroethene, vinyl chloride, ethene, and ethane.

Vinyl Chloride

Vinyl chloride is the organochloride with the formula $H_2C{:}CHCl$. It is also called vinyl chloride monomer, VCM or chloroethene. This colorless compound is an important industrial chemical chiefly used to produce the polymer polyvinyl chloride (PVC). At ambient pressure and temperature, vinyl chloride is a gas with a sickly sweet odor. It is highly toxic, flammable and carcinogenic.

History

Vinyl chloride was first produced in 1835 by Justus von Liebig and his student Henri Victor Regnault. They obtained it by treating ethylene dichloride with a solution of potassium hydroxide in ethanol.

In 1912, Frans, a German chemist working for Griesheim-Elektron, patented a means to produce vinyl chloride from acetylene and hydrogen chloride using mercuric chloride as a catalyst. Whereas this method was widely used during the 1930s and 1940s, it has since been superseded by more economical processes based on ethylene, at least in the West. It remains the main production in China.

Production

Vinyl chloride is produced on a substantial scale - approximately 31.1 million tons were produced in 2000. Two methods are employed, the hydrochlorination of acetylene and the dehydrochlorination of ethylene dichloride (1,2-dichloroethane).

Direct Chlorination

The production of vinyl chloride from 1,2-dichloroethane (EDC) consists of a series of well-defined steps. EDC is prepared by reacting ethylene and chlorine. In the presence of iron(III) chloride as a catalyst, these compounds react exothermically:

$$CH_2{=}CH_2 + Cl_2 \rightarrow ClCH_2CH_2Cl$$

This process is very selective, resulting in high purity EDC and high yields. However any dissolved catalyst and moisture must be removed before EDC enters the VCM production process.

Thermal Cracking

When heated to 500 °C at 15–30 atm (1.5 to 3 MPa) pressure, EDC vapour decomposes to produce vinyl chloride and anhydrous HCl.

$$ClCH_2CH_2Cl \rightarrow CH_2{=}CHCl + HCl$$

The thermal cracking reaction is highly endothermic, and is generally carried out in a fired heater. Even though residence time and temperature are carefully controlled, it produces significant quantities of chlorinated hydrocarbon side products. In practice, EDC conversion is relatively low (50 to 60 percent). The furnace effluent is immediately quenched with cold EDC to stop undesirable side reactions. The resulting vapour-liquid mixture then goes to a purification system. Some processes use an absorber-stripper system to separate HCl from the chlorinated hydrocarbons, while other processes use a refrigerated continuous distillation system.

Oxychlorination

Modern VCM plants use recycled HCl to produce more EDC via oxychlorination, which entails the reaction of ethylene, oxygen and hydrogen chloride over a copper(II) chloride catalyst to produce EDC:

$$CH_2{=}CH_2 + 2\ HCl + \tfrac{1}{2}\ O_2 \rightarrow ClCH_2CH_2Cl + H_2O.$$

The reaction is highly exothermic.

Due to the relatively low cost of ethylene, compared to acetylene, most vinyl chloride has been produced via this technique since the late 1950s. This is despite the lower yields, lower product purity and higher costs for waste treatment. By-products of the oxychlorination reaction, may be recovered, as feedstocks for chlorinated solvents production. One useful byproduct of the oxychlorination is ethyl chloride, a topical anesthetic.

Waste Treatment

For environmental reasons, the acidic aqueous wastestream is treated to remove organic compounds and neutralized before it can be sent to the plant's "outfall". An outfall is a monitored wastewater stream that must conform to the plant's standards. Some very hazardous wastes are generated in the recovery of the product vinyl chloride. These wastes require specialized procedures. These wastes are burned onsite in hazardous waste burners that again are subject to strict standards.

Production from Acetylene

Acetylene reacts with anhydrous hydrogen chloride gas over a mercuric chloride catalyst to give vinyl chloride:

$$C_2H_2 + HCl \rightarrow CH_2{=}CHCl$$

The reaction is exothermic and highly selective. Product purity and yields are generally very high.

This was the most common industrial route to VCM, before ethylene became widely distributed. When VCM producers shifted to using the thermal cracking of EDC described above, some used byproduct HCl in conjunction with a colocated acetylene-based unit. The hazards of storing and shipping acetylene meant that the VCM facility needed to be located very close to the acetylene generating facility.

Storage

Vinyl chloride is stored as a liquid. Often, the storage containers for the product VCM are high capacity spheres. The spheres have an inside sphere and an outside sphere. Several inches of empty space separate the inside sphere from the outside sphere. This void area between the spheres is purged with an inert gas such as nitrogen. As the nitrogen purge gas exits the void space it passes through an analyzer that is designed to detect if any vinyl chloride is leaking from the internal sphere. If vinyl chloride starts to leak from the internal sphere or if a fire is detected on the outside of the sphere then the contents of the sphere are automatically dumped into an emergency underground storage container.

Uses

Vinyl chloride is a chemical intermediate, not a final product. Due to the hazardous nature of vinyl chloride to human health there are no end products that use vinyl chloride in its monomer form. Polyvinyl chloride is very stable, storable, and nowhere near as acutely hazardous than the monomer.

Vinyl chloride liquid is fed to polymerization reactors where it is converted from a monomer to a polymer PVC. The final product of the polymerization process is PVC in either a flake or pellet form. From its flake or pellet form PVC is sold to companies that heat and mold the PVC into end products such as PVC pipe and bottles. Tens of billions of pounds of PVC are sold on the global market each year.

Until 1974, vinyl chloride was used in aerosol spray propellant. Prior to the removal of vinyl chloride from hair spray the accumulation of vinyl chloride vapour in hair salons may have exceeded the NOAEL (No Observable Adverse Effect Level) exposure guidelines.

Vinyl chloride was briefly used as an inhalational anaesthetic, in a similar vein to ethyl chloride, though its toxicity forced this practice to be abandoned.

Fire and Explosion Hazard

OSHA lists VCM as a Class IA Flammable Liquid, with an National Fire Protection Association Flammability Rating of 4. Because of its low boiling point, liquid VCM will undergo flash evaporation (i. e., autorefrigerate) upon its release to atmospheric pressure. The portion vaporized will form a dense cloud (more than twice as heavy as the surrounding air). The risk of subsequent explosion or fire is significant. According to OSHA, the flash point of VCM is -78 C (-108 F). Its flammable limits in air are: lower 3.6 volume % and upper 33.0 volume %. Fire may release toxic hydrogen chloride (HCl) and carbon monoxide (CO).

Health Effects

Vinyl Chloride Monomer

Almost all vinyl chloride monomer (VCM) is used to create polymers, primarily PVC (polyvinyl chloride).

Historically, workers in PVC plants were often exposed to high levels of VCM. Due to the carcinogenicity of VCM; many of those workers have contracted and died from cancer. The hepatotoxicity of VCM has long been established since the 1930s when the PVC industry was just in its infant stages. In the very first study about the dangers of Vinyl Chloride (VC), published by Patty in 1930, it was disclosed that exposure of test animals to just a single short-term high dose of VC caused liver damage. In 1949, a Russian publication by Tribukh discussed the finding that VC caused liver injury among workers. In

1954, Dr. Rex Wilson, Medical Director, and William McCormick, Industrial Hygienist and Toxicologist, both of B.F. Goodrich Chemical, published an article that stated that it was known VC caused liver injury for short-term exposures; but almost nothing was known about its long-term effects.

They also stated that long-term animal toxicology studies should be performed to fill this void of information. The study noted that if a chemical did not justify the cost of testing, and its ill-effects on workers and the public were known, the chemical should not be made. Thereafter, in 1963, Lester and Greenberg published an article reporting their findings from research paid for in part by Allied Chemical. They too found liver damage in test animals from exposures below 500 parts per million (ppm).

Then, in 1963, a Romanian researcher, Suciu, published his findings of liver disease in VC workers. In 1968, Mutchler and Kramer, two Dow researchers, reported their finding that exposures as low as 300 ppm caused liver damage in VC workers thus confirming earlier animal data in humans.

In a 1969 presentation given in Japan, Dr. P. L. Viola, a European researcher working for the European VC industry, indicated, "every monomer used in V.C. manufacture is hazardous....various changes were found in bone and liver. Particularly, much more attention should be drawn to liver changes.

The findings in rats at the concentration of 4 to 10 ppm are shown in pictures." In light of the finding of liver damage in rats from just 4-10 ppm of VC exposure, Dr. Viola added that he "should like some precautions to be taken in the manufacturing plants polymerizing vinyl chloride, such as a reduction of the threshold limit value of monomer ..." In 1970, Dr. Viola, reported that test animals exposed to 30,000 ppm of VC developed cancerous tumours. Viola began his research looking for the cause of liver and bone injuries found in VC workers. Dr. Viola's findings in 1970 were a "red flag" to B.F. Goodrich and the industry. In 1972, Dr. Maltoni, another Italian researcher for the European VC industry, found liver tumours (including angiosarcoma) from VC exposures as low as 250 ppm for four hours a day.

In the late 1960s the cancers that all of theses studies warned of finally manifested itself in workers. Dr. John Creech from

B.F. Goodrich discovered angiosarcoma (a very rare cancer) in the liver of a worker at the B.F. Goodrich plant in Louisville, Kentucky. Then, finally, on January 23, 1974, B.F. Goodrich informed the government and issued a press release stating that it was "investigating whether the cancer deaths of three employees in the polyvinyl chloride operations at its Louisville, Ky. plant were related to occupational causes." By then there really was no doubt that vinyl chloride caused angiosarcoma of the liver; it had been shown in both animal studies and worker experience.

A 1997 U.S. Centers for Disease Control and Prevention (CDC) report concluded that the development and acceptance by the PVC industry of a closed loop polymerization process in the late 1970s "almost completely eliminated worker exposures" and that "new cases of hepatic angiosarcoma in vinyl chloride polymerization workers have been virtually eliminated."

According to the United States Environmental Protection Agency (EPA), "vinyl chloride emissions from polyvinyl chloride (PVC), ethylene dichloride (EDC), and vinyl chloride monomer (VCM) plants cause or contribute to air pollution that may reasonably be anticipated to result in an increase in mortality or an increase in serious irreversible, or incapacitating reversible illness.

Vinyl chloride is a known human carcinogen that causes a rare cancer of the liver." EPA's 2001 updated Toxicological Profile and Summary Health Assessment for VCM in its Integrated Risk Information System (IRIS) database lowers EPA's previous risk factor estimate by a factor of 20 and concludes that "because of the consistent evidence for liver cancer in all the studies...and the weaker association for other sites, it is concluded that the liver is the most sensitive site, and protection against liver cancer will protect against possible cancer induction in other tissues."

A 1998 front-page series in the Houston Chronicle claimed the vinyl industry has manipulated vinyl chloride studies to avoid liability for worker exposure and to hide extensive and severe chemical spills into local communities.

Retesting of community residents in 2001 by the U.S. Agency for Toxic Substances and Disease Registry (ATSDR) found dioxin levels similar to those in a comparison community in Louisiana and to the U.S. population. Cancer rates in the community were similar to Louisiana and US averages.

Dioxins

The environmentalist group Greenpeace has advocated the global phase-out of PVC because they claim dioxin is produced as a byproduct of vinyl chloride manufacture and from incineration of waste PVC in domestic garbage. The European Industry, however, asserts that it has improved production processes to minimize dioxin emissions. Dioxins are a global health threat because they persist in the environment and can travel long distances. At very low levels, near those to which the general population is exposed, dioxins have been linked to immune system suppression, reproductive disorders, a variety of cancers, and endometriosis.

According to a 1994 report by the British firm, ICI Chemicals & Polymers Ltd., "It has been known since the publication of a paper in 1989 that these oxychlorination reactions [used to make vinyl chloride and some chlorinated solvents] generate polychlorinated dibenzodioxins (PCDDs) and dibenzofurans (PCDFs). The reactions include all of the ingredients and conditions necessary to form PCDD/PCDFs.... It is difficult to see how any of these conditions could be modified so as to prevent PCDD/PCDF formation without seriously impairing the reaction for which the process is designed." In other words, dioxins are an undesirable byproduct of polymerizing PVC and eliminating the production of dioxins while maintaining the polymerization reaction may be difficult. Dioxins created by vinyl chloride production are released by on-site incinerators, flares, boilers, wastewater treatment systems and even in trace quantities in vinyl resins. The US EPA estimate of dioxin releases from the PVC industry was 13 grams TEQ in 1995, or less than 0.5% of the total dioxin emissions in the US; by 2002, PVC industry dioxin emissions had been further reduced by 23%.

The largest well-quantified source of dioxin in the US EPA inventory of dioxin sources is barrel burning of household waste. Studies of household waste burning indicate consistent increases in dioxin generation with increasing PVC concentrations. According to the EPA dioxin inventory, landfill fires are likely to represent an even larger source of dioxin to the environment. A survey of international studies consistently identifies high dioxin concentrations in areas affected by open waste burning and a study that looked at the homologue pattern found the sample with the highest dioxin concentration was "typical for the pyrolysis of PVC". Other EU studies indicate that PVC likely "accounts for the overwhelming majority of chlorine that is available for dioxin formation during landfill fires."

The next largest sources of dioxin in the EPA inventory are medical and municipal waste incinerators. Studies have shown a clear correlation between dioxin formation and chloride content and indicate that PVC is a significant contributor to the formation of both dioxin and PCB in incinerators.

In February 2007, the Technical and Scientific Advisory Committee of the US Green Building Council (USGBC) released its report on a PVC avoidance related materials credit for the LEED Green Building Rating system. The report concludes that "no single material shows up as the best across all the human health and environmental impact categories, nor as the worst" but that the "risk of dioxin emissions puts PVC consistently among the worst materials for human health impacts."

Polyvinyl Chloride

Polyvinyl chloride, commonly abbreviated PVC, is a thermoplastic polymer. It is a vinyl polymer constructed of repeating vinyl groups (ethenyls) having one hydrogen replaced by chloride. Polyvinyl chloride is the third most widely produced plastic, after polyethylene and polypropylene. PVC is widely used in construction because it is cheap, durable, and easily worked. PVC production is expected to exceed 40 million tonnes by 2016. According to IUPAC, polyvinyl chloride should be named poly(chloroethanediyl), but this name found no widespread use.

It can be made softer and more flexible by the addition of plasticizers, the most widely used being phthalates. In this form, it is used in clothing and upholstery, electrical cable insulation, inflatable products and many applications in which it replaces rubber.

PVC is a controversial material, in part because of the generic term which is applied to many products, but each product may have a difference in the chemicals used during its production. For example, drinking water PVC pipes do not contain the controversial chemicals that household plastics contain. Other concerns extend to the products useful life and incineration, especially in accidental and uncontrolled circumstances, it may liberate persistent toxins, which the manufacture, use and destruction of suitable alternative plastics such as polypropylene do not.

Preparation

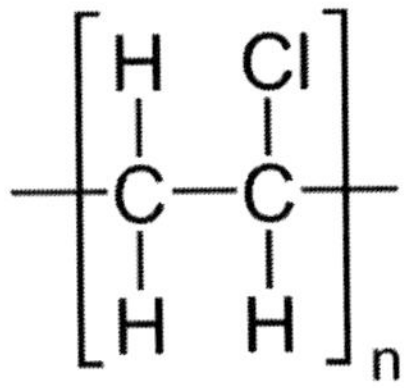

Space-filling model of a part of a PVC chain

Polyvinyl chloride is produced by polymerization of the vinyl chloride monomer (VCM), as shown. About 57% of its mass is chlorine.

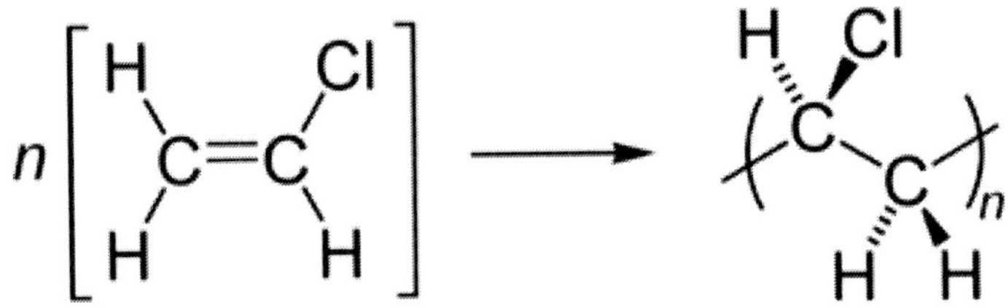

By far the most widely used production process is suspension polymerization. In this process, VCM and water are introduced into the polymerization reactor and a polymerization initiator, along with other chemical additives, are added to initiate the polymerization reaction. Usually a catalyst is also used for the better polymerization. The contents of the reaction vessel are continually mixed to maintain the suspension and ensure a uniform particle size of the PVC resin. The reaction is exothermic, and thus requires a cooling mechanism to maintain the reactor contents at the appropriate temperature. As the volumes also contract during the reaction (PVC is denser than VCM), water is continually added to the mixture to maintain the suspension.

Once the reaction has run its course, the resulting PVC slurry is degassed and stripped to remove excess VCM (which is recycled) then passed though a centrifuge to remove water. The slurry is further dried in a hot air bed, and the resulting powder sieved before storage or pelletization. Normally, the resulting PVC has a VCM content of less than 1 part per million.

Other production processes, such as micro-suspension polymerization and emulsion polymerization, produce PVC with smaller particle sizes (10 μm vs. 120–150 μm for suspension PVC) with slightly different properties and with somewhat different sets of applications. The product of the polymerization process is unmodified PVC.

Before PVC can be made into finished products, it almost always requires conversion into a compound by the incorporation of additives such as heat stabilizers, UV stabilizers, lubricants, plasticizers, processing aids, impact modifiers, thermal modifiers, fillers, flame retardants, biocides, blowing agents and smoke suppressors, and, optionally pigments.

Properties

The properties for PVC are usually categorized based on rigid and flexible PVCs.

Property	*Rigid PVC*	*Flexible PVC*
Density [g/cm^3]	1.3–1.45	1.1–1.35
Thermal conductivity [W/(m·K)]	0.14–0.28	0.14–0.17
Yield strength [MPa]	31–60	10–25
Young's modulus [psi]	490,000	
Flexural strength (yield) [psi]	10,500	
Compression strength [psi]	9500	
Coefficient of thermal expansion (linear) [mm/(mm °C)]	5×10^{-5}	
Vicat B [°C]	65–100	Not recommended
Resistivity [Ω m]	10^{16}	10^{12}–10^{15}
Surface resistivity [Ω]	10^{13}–10^{14}	10^{11}–10^{12}

History

PVC was accidentally discovered at least twice in the 19th century, first in 1835 by Henri Victor Regnault and in 1872 by Eugen Baumann. On both occasions the polymer appeared as a white solid inside flasks of vinyl chloride that had been left exposed to sunlight.

In the early 20th century the Russian chemist Ivan Ostromislensky and Fritz Klatte of the German chemical company Griesheim-Elektron both attempted to use PVC in commercial products, but difficulties in processing the rigid, sometimes brittle polymer blocked their efforts. Waldo Semon and the B.F. Goodrich Company developed a method in 1926 to plasticize PVC by blending it with various additives. The result was a more flexible and more easily processed material that soon achieved widespread commercial use.

Applications

A number of PVC's properties recommend it for a wide variety of applications. It is biologically and chemically resistant, making it the plastic of choice for most household sewerage pipes and other pipe applications where corrosion would limit the use of metal. It is also a suitable material for greenhouses, small home playgrounds, furniture, pool toy frames, foam weapons, custom truck toppers, or even giant building sets for kids.

With the addition of impact modifiers and stabilizers, it becomes a popular material for window and door frames. By adding plasticizers, it can become flexible enough to be used in cabling applications as a wire insulator. It is also used to make vinyl records.

Clothing

PVC has become widely used in clothing, to either create a leather-like material or at times simply for the effect of PVC. PVC clothing is common in Goth, Punk and alternative fashions. PVC is cheaper than rubber, leather, and latex and so it is more widely available and worn.

PVC fabric has a sheen to it and is waterproof. It is commonly used in coats, skiing equipment, shoes, jackets, aprons, and bags because of this.

Electric Wires

PVC is commonly used as the insulation on electric wires; the plastic used for this purpose needs to be plasticized.

In a fire, PVC-coated wires can form HCl fumes; the chlorine serves to scavenge free radicals and is the source of the material's fire retardance.

While HCl fumes can also pose a health hazard in their own right, HCl dissolves in moisture and breaks down onto surfaces, particularly in areas where the air is cool enough to breathe, and is not available for inhalation. Frequently in applications where smoke is a major hazard (notably in tunnels and communal areas) PVC-free cable insulation is preferred, such as low smoke zero halogen (LSZH) insulation.

Pipes

Roughly half of the world's polyvinyl chloride resin manufactured annually is used for producing pipes for various municipal and industrial applications. In the water distribution market it accounts for 66% of the market in the US, and in sanitary sewer pipe applications, it accounts for 75%. Its light weight, high strength, and low reactivity make it particularly well-suited to this purpose. In addition, PVC pipes can be fused together using various solvent cements, or heat-fused (butt-fusion process, similar to joining HDPE pipe), creating permanent joints that are virtually impervious to leakage.

In February, 2007 the California Building Standards Code was updated to approve the use of chlorinated polyvinyl chloride (CPVC) pipe for use in residential water supply piping systems. CPVC has been a nationally accepted material in the US since 1982; California, however, has permitted only limited use since 2001. The Department of Housing and Community Development prepared and certified an environmental impact statement resulting in a recommendation that the Commission adopt and approve the use of CPVC. The Commission's vote was unanimous and CPVC has been placed in the 2007 California Plumbing Code.

In the United States and Canada, PVC pipes account for the largest majority of pipe materials used in buried municipal applications for drinking water distribution and wastewater mains.

Signs

Polyvinyl chloride is formed in flat sheets in a variety of thicknesses and colours. As flat sheets, PVC is often expanded to create voids in the interior of the material, providing additional thickness without additional weight and minimal extra cost. Sheets are cut using saw and rotary cutting equipment. Plasticized PVC is also used to produce thin, coloured, or clear, adhesive-backed films referred to simply as vinyl. These films are typically cut on a computer-controlled plotter or printed in a wide-format printer. These sheets and films are used to produce a wide variety of commercial signage products and markings on vehicles, e.g. car body stripes.

Joining

PVC cements are available at plumbing supply houses. The cement softens the material to a gel state until the adhesive layer cures. This has another practical application of being able to hand-machine PVC pipe with a razor blade to change the wall thickness to allow assembly of nonstandard radius arc segments of electrical conduit large radius elbows. Buried PVC pipes in both water and sanitary sewer applications that are 4 inches (100 mm) in diameter and larger are typically joined by means of a gasket-sealed joint. In gasket-sealed jointing systems, an elastomeric gasket is seated in a groove within the bell (female end of pipe). When the spigot (male end) of the adjoining pipe is inserted into the bell, the gasket forms a positive seal that prevents leakage of fluid within the pipe and also keeps water and other contaminants from surrounding soils from entering the piping system. The most common type of gasket utilized in North America is a metal reinforced elastomer, commonly referred to as a Reiber sealing system.

Ceiling Tiles

PVC ceiling tiles are an alternative ceiling tiles that are easy to install over any flat surface. They can be glued onto an existing ceiling with a reasonably flat surface.

Unplasticized Polyvinyl Chloride (uPVC)

uPVC or rigid PVC is often used in the building industry as a low-maintenance material, particularly in Ireland, the United Kingdom, and in the United States where it is known as vinyl, or vinyl siding. The material comes in a range of colours and finishes, including a photo-effect wood finish, and is used as a substitute for painted wood, mostly for window frames and sills when installing double glazing in new buildings, or to replace older single glazed windows. It has many other uses including fascia, and siding or weatherboarding. The same material has almost entirely replaced the use of cast iron for plumbing and drainage, being used for waste pipes, drainpipes, gutters and downspouts. uPVC does not contain phthalates or BPA. Most dental retainers and mouthguards are made from uPVC. uPVC does not have the same concerns as flexible PVC. Phthalates are only added to flexible PVC. uPVC is also known as rigid PVC, uPVC is known as having strong resistance against chemicals, sunlight, and oxidation from water. uPVC stands for un-plasticized PVC.

Health and Safety

Phthalate Plasticizers

Many vinyl products contain additional chemicals to change the chemical consistency of the product. Some of these additional chemicals called additives can leach out of vinyl products. Plasticizers that must be added to make PVC flexible have been additives of particular concern.

Because soft PVC toys have been made for babies for years, there are concerns that these additives leach out of soft toys into the mouths of the children chewing on them. Phthalates mimic human hormones and also affect various life forms including fish and invertebrates adversely. Additionally, adult sex toys have been demonstrated to contain high concentrations of the additives. In January 2006, the European Union placed a ban on six types of phthalate softeners, including DEHP (diethylhexyl phthalate), used in toys. In the U.S. most companies have voluntarily stopped manufacturing PVC toys with DEHP and in 2003 the U.S. Consumer Product Safety Commission (CPSC) denied a petition for a ban on PVC toys made with an alternative plasticizer, DINP (diisononyl phthalate). In April 2006, the European Chemicals Bureau of the European Commission published an assessment of DINP which found risk "unlikely" for children and newborns.

Vinyl intravenous (IV) bags used in neo-natal intensive care units have also been shown to leach DEHP. In a draft guidance paper published in September 2002, the US FDA recognizes that many medical devices with PVC containing DEHP are not used in ways that result in significant human exposure to the chemical. The FDA, however, suggests that manufacturers consider eliminating DEHP in certain devices that can result in high aggregate exposures for sensitive patient populations such as neonates. Due to environmental concerns use of PVC is discouraged by some local authorities in countries such as UK, Germany and the Netherlands.

Other vinyl products including car interiors, shower curtains, and flooring initially release chemical gases into the air. Some studies indicate that this outgassing of additives may contribute to health complications, and have resulted in a call for banning the use of DEHP on shower curtains, among other uses. The Japanese car companies Toyota, Nissan, and Honda have eliminated PVC in their car interiors starting in 2007.

In 2004 a joint Swedish-Danish research team found a statistical association between allergies in children and indoor air levels of DEHP and BBzP (butyl benzyl phthalate), which is used in vinyl flooring. In December 2006, the European Chemicals Bureau of the European Commission released a final draft risk assessment of BBzP which found "no concern" for consumer exposure including exposure to children.

In November, 2005 one of the largest hospital networks in the U.S., Catholic Healthcare West, signed a contract with B.Braun for vinyl-free intravenous bags and tubing. According to the Centre for Health, Environment & Justice in Falls Church, VA, which helps to coordinate a "precautionary" PVC Campaign, several major corporations including Microsoft, Wal-Mart, and Kaiser Permanente announced efforts to eliminate PVC from products and packaging in 2005. Target is reducing its sale of items with PVC.

The FDA Paper titled "Safety Assessment of Di(2-ethylhexyl)phthalate (DEHP)Released from PVC Medical Devices" states that Critically ill or injured patients may be at increased risk of developing adverse health effects from DEHP, not only by virtue of increased exposure, relative to the general population, but also because of the physiological and pharmacodynamic changes that occur in these patients, compared to healthy individuals.

In 2008 the European Union's Scientific Committee on Emerging and Newly Identified Health Risks (SCENIHR) reviewed the safety of DEHP in medical devices. The SCENIHR report states that certain medical procedures used in high risk patients result in a significant exposure to DEHP and concludes there is still a reason for having some concerns about the exposure of prematurely born male babies to medical devices containing DEHP. The Committee said there are some alternative plasticizers available for which there is sufficient toxicological data to indicate a lower hazard compared to DEHP but added that the functionality of these plasticizers should be assessed before they can be used as an alternative for DEHP in PVC medical devices.

Vinyl Chloride Monomer

In the early 1970s, Dr. John Creech and Dr. Maurice Johnson were the first to clearly link and recognize the carcinogenicity of vinyl chloride monomer to humans when workers in the polyvinyl chloride polymerization section of a B.F. Goodrich plant near Louisville, Kentucky, were diagnosed with liver angiosarcoma also known as hemangiosarcoma, a rare disease. Since that time, studies of PVC workers in Australia, Italy, Germany, and the UK have all associated certain types of occupational cancers with exposure to vinyl chloride.

The link between angiosarcoma of the liver and long-term exposure to vinyl chloride is the only one that has been confirmed by the International Agency for Research on Cancer. All the cases of angiosarcoma developed from exposure to vinyl chloride monomer were in workers who were exposed to very high VCM levels, routinely, for many years. These workers cleaned accretions in reactors, a practice that has now been replaced by automated high-pressure water jets.

A 1997 U.S. Centers for Disease Control and Prevention (CDC) report concluded that the development and acceptance by the PVC industry of a closed loop polymerization process in the late 1970s "almost completely eliminated worker exposures" and that "new cases of hepatic angiosarcoma in vinyl chloride polymerization workers have been virtually eliminated."

According to the EPA, "vinyl chloride emissions from polyvinyl chloride (PVC), ethylene dichloride (EDC), and vinyl chloride monomer (VCM) plants cause or contribute to air pollution that may reasonably be anticipated to result in an increase in mortality or an increase in serious irreversible, or incapacitating reversible illness.

Vinyl chloride is a known human carcinogen that causes a rare cancer of the liver." EPA's 2001 updated Toxicological Profile and Summary Health Assessment for VCM in its Integrated Risk Information System (IRIS) database lowers EPA's previous risk factor estimate by a factor of 20 and concludes that "because of the consistent evidence for liver cancer in all the studies...and the weaker association for other sites, it is concluded that the liver is the most sensitive site, and protection against liver cancer will protect against possible cancer induction in other tissues."

A 1998 front-page series in the Houston Chronicle claimed the vinyl industry has manipulated vinyl chloride studies to avoid liability for worker exposure and to hide extensive and severe chemical spills into local communities. Retesting of community residents in 2001 by the U.S. Agency for Toxic Substances and Disease Registry (ATSDR) found dioxin levels similar to those in a comparison community in Louisiana and to the U.S. population. Cancer rates in the community were similar to Louisiana and US averages.

Bans

The State of California is currently considering a bill that would ban the use of PVC in consumer packaging due to the threats it poses to human and environmental health and its effect on the recycling stream. Specifically, the language of the bill analysis stipulates that EPA has listed vinyl chloride, a "constituent element" of PVC, as a carcinogen. It also further cites that there are concerns about the leaching of phthalates and lead from the PVC packaging.

Recycling

In Europe, developments in PVC waste management are monitored by Vinyl 2010, a legal entity established in 2000. In the waste management area their commitment is to

1. Support integrated waste management approaches, using raw materials as efficiently as possible
2. Work with the stakeholders to research, develop, and implement recycling of 200,000 tonnes per year of PVC postconsumer waste in 2010 in addition to waste already recycled in 2000, or regulated by the PPW, ELV and E&E Waste Directives
3. Recycle at least 25% of collectable, available PVC postconsumer waste from pipes, window profiles, and roofing membranes by 2003 and increase this to 50% by 2005.

Vinyl 2010 has a Monitoring Committee and publishes annual reviews. In 2011, it reported that 260,842 tonnes of post-consumer PVC waste was recycled in 2010, i.e. an increase of 220,000 tonnes over the 1999 volumes, exceeding the 10-year target of 200,000 tonnes.

Collection and recycling schemes for PVC waste stream are managed through Recovinyl which reported the recycled tonnage as follows: pipe 25 kT, profile 107 kT, rigid film 6 kT, flexible cables 79 kt and mixed flexible 38 kT.

One of the recycling processes is the Vinyloop Texyloop used for solvent-based mechanical recycling. It involves recovering PVC plastic from composite materials through dissolution and precipitation, and is a closed-loop system, recycling the solvent and regenerating PVC.

Propene

Propene, also known as propylene or methylethylene, is an unsaturated organic compound having the chemical formula C_3H_6. It has one double bond, and is the second simplest member of the alkene class of hydrocarbons, and it is also second in natural abundance.

Properties

At room temperature and atmospheric pressure, propene is a gas, and as with many other alkenes, it is also colourless with a weak but unpleasant smell.

Propene has a higher density and boiling point than ethylene due to its greater size. It has a slightly lower boiling point than propane and is thus more volatile. It lacks strongly polar bonds, yet the molecule has a small dipole moment due to its reduced symmetry (its point group is C_s).

Propene has the same empirical formula as cyclopropane but their atoms are connected in different ways, making these molecules structural isomers.

Production

Propene is produced from fossil fuels—petroleum, natural gas, and, to a much lesser extent, coal. Propene is a byproduct of oil refining and natural gas processing. During oil refining, ethylene, propene, and other compounds are produced by as a result of cracking larger hydrocarbon molecules to produce hydrocarbons more in demand.

A major source of propene is cracking intended to produce ethylene, but it also results from refinery cracking producing other products. Propene can be separated by fractional distillation from hydrocarbon mixtures obtained from cracking and other refining processes; refinery-grade propene is about 50 to 70%.

Another important petrochemical source of propene is propane dehydrogenation. This route is popular in regions, such as the Middle East, where there is an abundance of propane from oil/gas operations. Less common propene sources are the Fischer-Tropsch process, metathesis of ethylene or an ethylene/butene mixture, and catalytic conversion of methanol.

Propene production has remained static at around 35 million tonnes (Europe and North America only) from 2000 to 2008, but it has been increasing in East Asia, most notably Singapore and China. Total world production of propene is currently about half that of ethylene.

Uses

Propene is the second most important starting product in the petrochemical industry after ethylene. It is the raw material for a wide variety of products. Manufacturers of the plastic polypropylene account for nearly two thirds of all demand. Polypropylene is, for example, needed for the production of films, packaging, caps and closures as well as for other applications. In the year 2008 the worldwide sales of propene reached a value of over 90 billion US dollars.

Propene and benzene are converted to acetone and phenol via the cumene process. Propene is also used to produce isopropanol (propan-2-ol), acrylonitrile, propylene oxide (epoxypropane) and epichlorohydrin.

Reactions

Propene resembles other alkenes in that it undergoes addition reactions relatively easily at room temperature. The relative weakness of its double bond (which is less strong than two single bonds) explains its tendency to react with substances that can achieve this transformation. Alkene reactions include: 1) polymerization, 2) oxidation, 3) halogenation and hydrohalogenation, 4) alkylation, 5) hydration, 6) oligomerization, and 7) hydroformylation.

Environmental Safety

Propene is produced naturally by vegetation, particularly certain tree species. It is also a product of combustion, from forest fires and cigarette smoke to motor vehicle and aircraft exhaust. It is an impurity in some heating gases. Observed concentrations have been in the range of 0.1-4.8 parts per billion (ppb) in rural air, 4-10.5 ppb in urban air, and 7-260 ppb in industrial air samples.

In the United States and some European countries a Threshold Limit Value of 500 parts per million (ppm) was established for occupational (8-hour time-weighted average) exposure. It is considered a volatile organic compound (VOC) and emissions are regulated by many governments, but it is not listed by the U.S. Environmental Protection Agency (EPA) as a hazardous air pollutant under the Clean Air Act. It has a relatively short half-life in the atmosphere, and is not expected to bioaccumulate, based on a calculated bioconcentration factor of 13.18 using a log K_{ow} value of 1.77.

Propene has low acute toxicity from inhalation. Inhalation of the gas can cause anesthetic effects and at very high concentrations, unconsciousness. However, the asphyxiation limit for humans is about 10 times higher (23%) than the lower flammability level.

Isopropyl Alcohol

Isopropyl alcohol (also Isopropanol, propan-2-ol, 2-propanol or the abbreviation IPA) is a common name for a chemical compound with the molecular formula C_3H_8O. It is a colorless, flammable chemical compound with a strong odor. It is the simplest example of a *secondary alcohol*, where the alcohol carbon is attached to two other carbons sometimes shown as $(CH_3)_2CHOH$. It is a structural isomer of propanol.

Properties

Isopropyl alcohol has a maximum absorbance at 204 nm in an ultraviolet-visible spectrum. Unlike ethanol or methanol, isopropyl alcohol can be separated from aqueous solutions by adding a salt such as sodium chloride, sodium sulfate, or any of several other inorganic salts, since the alcohol is much less soluble in saline solutions than in salt-free water The process is colloquially called salting out, and causes concentrated isopropyl alcohol to separate into a distinct layer.

Being a secondary alcohol, isopropyl alcohol can be oxidized to acetone, which is the corresponding ketone. This can be achieved using oxidizing agents such as chromic acid, or by dehydrogenation of isopropyl alcohol over a heated copper catalyst:

$$(CH_3)_2CH\text{-}OH \rightarrow (CH_3)_2CO + H_2$$

Isopropyl alcohol is often used as both solvent and hydride source in the Meerwein-Ponndorf-Verley reduction and other transfer hydrogenation reactions; it is oxidized to acetone.

Isopropyl alcohol may be converted to 2-bromopropane using phosphorus tribromide, or dehydrated to propene by heating with sulfuric acid.

Like most alcohols, isopropyl alcohol reacts with active metals such as potassium to form alkoxides which can be called *isopropoxides*. The reaction with aluminium (initiated by a trace of mercury) is used to prepare the catalyst aluminium isopropoxide.

Production

In 1994, 1.5 million tonnes of isopropyl alcohol was produced in the United States, Europe, and Japan. This compound is primarily produced by combining water and propene in a hydration reaction.

Of minor significance is the hydrogenation of acetone. There are two routes for the hydration process: indirect hydration via the sulfuric acid process, and direct hydration. The former process, which can use low-quality propene, predominates in the USA while the latter process, which requires high-purity propene, is more commonly used in Europe. These processes give predominantly isopropyl alcohol rather than 1-propanol because the addition of water or sulfuric acid to propene follows Markovnikov's rule.

Indirect Hydration

The indirect process reacts propene with sulfuric acid to form a mixture of sulfate esters. Subsequent hydrolysis of these esters by steam produces isopropyl alcohol, which is distilled. Diisopropyl ether is a significant by-product of this process; it is recycled back to the process and hydrolyzed to give the desired product.

Direct Hydration

Direct hydration reacts propene and water, either in gas or liquid phases, at high pressures in the presence of solid or supported acidic catalysts. Higher purity propylene (> 90 %) tends to be required for this type of process.

Both processes require that the isopropyl alcohol be separated from water and other by-products by distillation. Isopropyl alcohol and water form an azeotrope and simple distillation gives a material which is 87.9% by weight isopropyl alcohol and 12.1% by weight water. Pure (anhydrous) isopropyl alcohol is made by azeotropic distillation of the wet isopropyl alcohol using either diisopropyl ether or cyclohexane as azeotroping agents.

Hydrogenation of Acetone

Crude acetone is hydrogenated in the liquid phase over Raney nickel or a mixture of copper and chromium oxide to give isopropyl alcohol. This process is useful when coupled with excess acetone production, such as the cumene process.

Uses

In 1990, 45 thousand tonnes of isopropyl alcohol was consumed in the United States. The vast majority of isopropyl alcohol was used as a solvent for coatings or for industrial processes. Isopropyl alcohol in particular is popular for pharmaceutical applications, presumably due to the low toxicity of any residues. Some isopropyl alcohol is used as a chemical intermediate. Isopropyl alcohol may be converted to acetone, but the cumene process is more significant. In that year, a significant fraction (5.4 tonnes) was consumed for household use and in personal care products. It is also used as a gasoline additive.

Solvent

Isopropyl alcohol dissolves a wide range of non-polar compounds. It also evaporates quickly and is relatively non-toxic, compared to alternative solvents. Thus it is used widely as a solvent and as a cleaning fluid, especially for dissolving oils.

Examples of this application include cleaning electronic devices such as contact pins (like those on ROM cartridges), magnetic tape and disk heads (such as those in audio and video tape recorders and floppy disk drives), the lenses of lasers in optical disc drives (e.g. CD, DVD) and removing thermal paste from heatsinks and IC packages (such as CPUs.)

Isopropyl alcohol is used in keyboard, LCD and laptop cleaning, is sold commercially as a whiteboard cleaner, and is a strong but safer alternative to common household cleaning products. It is used to clean LCD and glass computer monitor screens (at some risk to the anti-reflection coating on some screens), and used to give second-hand or worn non-vinyl phonograph records newer-looking sheen.

Isopropyl alcohol should not be used to clean vinyl records as it may leach plasticizer from the vinyl making it more rigid). It is effective at removing residual glue from some sticky labels although some other adhesives used on tapes and paper labels are resistant to it. It can also be used to remove stains from most fabrics, wood, cotton, etc. In addition it can also be used to clean paintballs or other oil based products so that they may be reused, commonly known as “repainting”.

Intermediate

Isopropyl alcohol is esterified to give isopropyl acetate, another solvent. It reacts with carbon disulfide to give sodium isopropylxanthate, a weed killer. Isopropyl alcohol reacts with titanium tetrachloride and aluminium metal to give titanium and aluminium isopropoxides respectively, the former a catalyst, and the latter a chemical reagent. This compound may serve as a chemical reagent in itself, by acting as a dihydrogen donor in transfer hydrogenation.

Medical

Disinfecting pads typically contain a 60–70% solution of isopropyl alcohol in water. A 75 % v/v solution in water may be used as a hand sanitizer. Isopropyl alcohol is used as a water-drying aid for the prevention of otitis externa, better known as swimmer's ear.

Automotive

Isopropyl alcohol is a major ingredient in "gas dryer" fuel additives. In significant quantities, water is a problem in fuel tanks, as it separates from the gasoline, and can freeze in the supply lines at cold temperatures.

It does not remove water from gasoline; rather, the alcohol solubilizes water in gasoline. Once soluble, water does not pose the same risk as insoluble water as it will no longer accumulate in the supply lines and freeze. Isopropyl alcohol is often sold in aerosol cans as a windscreen de-icer. Isopropyl alcohol is also used to remove brake fluid traces from hydraulic braking systems, so that the brake fluid (usually DOT 3, DOT 4 or mineral oil) does not contaminate the brake pads, which would result in poor braking.

Laboratory

As a biological specimen preservative, isopropyl alcohol provides a comparatively non-toxic alternative to formaldehyde and other synthetic preservatives. Isopropyl alcohol solutions of 90-99% are optimal for preserving specimens, although concentrations as low as 70% can be used in emergencies.

Isopropyl alcohol is often used in DNA extraction. It is added to a DNA solution in order to precipitate the DNA into a 'pellet' after centrifuging the DNA. This is possible because DNA is insoluble in isopropyl alcohol.

Safety

Isopropyl alcohol vapour is denser than air and is highly flammable with a very wide combustible range. It should be kept away from heat and open flame. When mixed with air or other oxidizers it can explode through deflagration. Isopropyl alcohol has also been reported to form explosive peroxides.

Like many organic solvents, long term application to the skin can cause defatting.

Toxicology

Isopropyl alcohol is oxidized by the liver into acetone by alcohol dehydrogenase. Symptoms of isopropyl alcohol poisoning include flushing, headache, dizziness, CNS depression, nausea, vomiting, anesthesia, and coma. Poisoning can occur from ingestion, inhalation, or absorption; therefore, well-ventilated areas and protective gloves are recommended.

Isopropyl alcohol and its metabolite, acetone, act as central nervous system (CNS) depressants. Around 15 g of isopropyl alcohol can have a toxic effect on a 70 kg human if left untreated. However, it is not nearly as toxic as methanol or ethylene glycol. Isopropyl alcohol does not cause an anion gap acidosis (in which as lowered blood serum pH causes depletion of bicarbonate anion) unlike ethanol and methanol. Isopropyl alcohol does however, produce an osmolal gap between the calculated and measured osmolalities of serum, as do the other alcohols. Overdoses may cause a fruity odor on the breath as a result of its metabolism to acetone, which is further metabolized to produce the nutrients acetate and glucose.

Acrylonitrile

Acrylonitrile is the chemical compound with the formula C_3H_3N. This pungent-smelling colorless liquid often appears yellow due to impurities. It is an important monomer for the manufacture of useful plastics. In terms of its molecular structure, it consists of a vinyl group linked to a nitrile. Pathways of exposure include emissions, auto exhaust, and cigarette smoke that can expose the human subject directly if they inhale or smoke. Routes of exposure include inhalation, oral, and occasional dermal routes from volunteer humans and rat studies.

Uses

Acrylonitrile is used principally as a monomer to prepare the polyacrylonitrile, a homopolymer, or several important copolymers such as styrene-acrylonitrile (SAN), acrylonitrile butadiene styrene (ABS), acrylonitrile styrene acrylate (ASA) and other synthetic rubbers such as acrylonitrile butadiene (NBR). Dimerization of acrylonitrile affords adiponitrile, used in the synthesis of certain polyamides. Small amounts are also used as a fumigant. Acrylonitrile and derivatives such as 2-chloro-acrylonitrile are dienophiles in Diels-Alder reactions.

Acrylonitrile is also a precursor in the industrial manufacture of acrylamide and acrylic acid.

Production

Of all the nitriles, Acrylonitrile is manufactured on probably the largest scale. Most industrial acrylonitrile is produced by catalytic ammoxidation of propylene:

$$2CH_3\text{-}CH{=}CH_2 + 2NH_3 + 3O_2 \rightarrow 2CH_2{=}CH\text{-}Ca\text{–}N + 6H_2O$$

Health Effects

Acrylonitrile is highly flammable and toxic. It undergoes explosive polymerization. The burning material releases fumes of hydrogen cyanide and oxides of nitrogen. The International Agency for Research on Cancer (IARC) concluded that there is inadequate evidence in humans for the carcinogenicity of acrylonitrile, but classified it as a Class 2B carcinogen (possibly carcinogenic). Acrylonitrile increases cancer in high dose tests in male and female rats and mice.

There are two main excretion processes of acrylonitrile. The primary method is excretion in urine when acrylonitrile is metabolized by being directly conjugated to glutathione. The other method is when acrylonitrile is metabolized with 2-cyanoethylene oxide to produce cyanide end products that ultimately forms thiocyanate, which is excreted via urine, or carbon dioxide and eliminated through the lungs.

Polypropylene

Polypropylene (PP), also known as polypropene, is a thermoplastic polymer used in a wide variety of applications including packaging, textiles (e.g., ropes, thermal underwear and carpets), stationery, plastic parts and reusable containers of various types, laboratory equipment, loudspeakers, automotive components, and polymer banknotes. An addition polymer made from the monomer propylene, it is rugged and unusually resistant to many chemical solvents, bases and acids.

In 2007, the global market for polypropylene had a volume of 45.1 million tons, which led to a turnover of about \$65 billion (~ €47.4 billion).

Chemical and Physical Properties

Most commercial polypropylene is isotactic and has an intermediate level of crystallinity between that of low-density polyethylene (LDPE) and high-density polyethylene (HDPE). Polypropylene is normally tough and flexible, especially when copolymerized with ethylene. This allows polypropylene to be used as an engineering plastic, competing with materials such as ABS. Polypropylene is reasonably economical, and can be made translucent when uncolored but is not as readily made transparent as polystyrene, acrylic, or certain other plastics. It is often opaque or coloured using pigments. Polypropylene has good resistance to fatigue.

The melting of polypropylene occurs as a range, so a melting point is determined by finding the highest temperature of a differential scanning calorimetry chart. Perfectly isotactic PP has a melting point of 171 °C (340 °F). Commercial isotactic PP has a melting point that ranges from 160 to 166 °C (320 to 331 °F), depending on atactic material and crystallinity. Syndiotactic PP with a crystallinity of 30% has a melting point of 130 °C (266 °F).

The melt flow rate (MFR) or melt flow index (MFI) is a measure of molecular weight of polypropylene. The measure helps to determine how easily the molten raw material will flow during processing. Polypropylene with higher MFR will fill the plastic mold more easily during the injection or blow-molding production process. As the melt flow increases, however, some physical properties, like impact strength, will decrease.

There are three general types of polypropylene: homopolymer, random copolymer, and block copolymer. The comonomer is typically used with ethylene. Ethylene-propylene rubber or EPDM added to polypropylene homopolymer increases its low temperature impact strength. Randomly polymerized ethylene monomer added to polypropylene homopolymer decreases the polymer crystallinity and makes the polymer more transparent.

Degradation

Polypropylene is liable to chain degradation from exposure to heat and UV radiation such as that present in sunlight. Oxidation usually occurs at the tertiary carbon atom present in every repeat unit. A free radical is formed here, and then reacts further with oxygen, followed by chain scission to yield aldehydes and carboxylic acids. In external applications, it shows up as a network of fine cracks and crazes that become deeper and more severe with time of exposure.

For external applications, UV-absorbing additives must be used. Carbon black also provides some protection from UV attack. The polymer can also be oxidized at high temperatures, a common problem during molding operations. Anti-oxidants are normally added to prevent polymer degradation.

History

Propylene was first polymerized to a crystalline isotactic polymer by Giulio Natta as well as by the German chemist Karl Rehn in March 1954. This pioneering discovery led to large-scale commercial production of isotactic polypropylene from 1957 onwards. Syndiotactic polypropylene was also first synthesized by Giulio Natta and his coworkers.

Synthesis

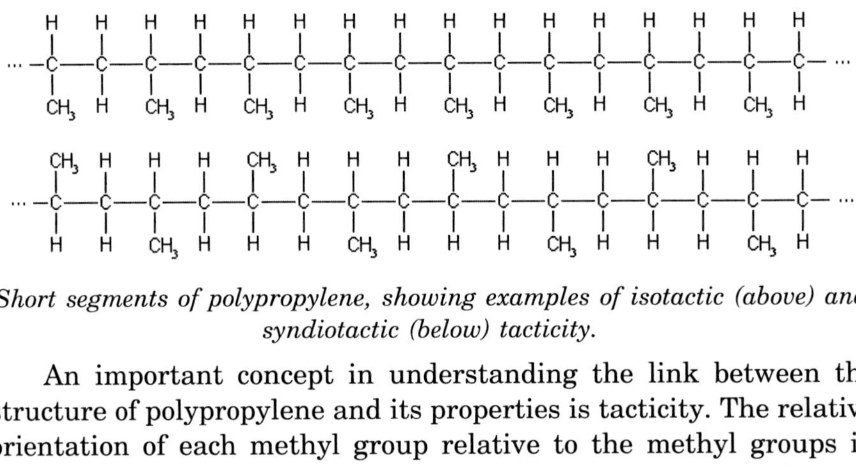

Short segments of polypropylene, showing examples of isotactic (above) and syndiotactic (below) tacticity.

An important concept in understanding the link between the structure of polypropylene and its properties is tacticity. The relative orientation of each methyl group relative to the methyl groups in neighboring monomer units has a strong effect on the polymer's ability to form crystals.

A Ziegler-Natta catalyst is able to restrict linking of monomer molecules to a specific regular orientation, either isotactic, when all methyl groups are positioned at the same side with respect to the backbone of the polymer chain, or syndiotactic, when the positions of the methyl groups alternate. Commercially available isotactic polypropylene is made with two types of Ziegler-Natta catalysts. The first group of the catalysts encompases solid (mostly supported) catalysts and certain types of soluble metallocene catalysts. Such isotactic macromolecules coil into a helical shape; these helices then line up next to one another to form the crystals that give commercial isotactic polypropylene many of its desirable properties. Another type of metallocene catalysts produce syndiotactic polypropylene. These macromolecules also coil into helices (of a different type) and form crystalline materials.

When the methyl groups in a polypropylene chain exhibit no preferred orientation, the polymers are called atactic. Atactic polypropylene is an amorphous rubbery material. It can be produced commercially either with a special type of supported Ziegler-Natta catalyst or with some metallocene catalysts.

Modern supported Ziegler-Natta catalysts developed for the polymerization of propylene and other 1-alkenes to isotactic polymers usually use $TiCl_4$ as an active ingredient and $MgCl_2$ as a support.,, The catalysts also contain organic modifiers, either aromatic acid esters and diesters or ethers. These catalysts are activated with special cocatalysts containing an organoaluminum compound such as $Al(C_2H_5)_3$ and the second type of a modifier. The catalysts are differentiated depending on the procedure used for fashioning catalyst particles from $MgCl_2$ and depending on the type of organic modifiers employed during catalyst preparation and use in polymerization reactions. Two most important technological characteristics of all the supported catalysts are high productivity and a high fraction of the crystalline isotactic polymer they produce at 70-80 °C under standard polymerization conditions. Commercial synthesis of isotactic polypropylene is usually carried out either in the medium of liquid propylene or in gas-phase reactors.

Commercial synthesis of syndiotactic polypropylene is carried out with the use of a special class of metallocene catalysts. They employ bridged bis-metallocene complexes of the type bridge-$(Cp_1)(Cp_2)ZrCl_2$ where the first Cp ligand is the cyclopentadienyl group, the second Cp ligand is the fluorenyl group, and the bridge between the two Cp ligands is $-CH_2-CH_2-$, $>SiMe_2$, or $>SiPh_2$. These complexes are converted to polymerization catalysts by activating them with a special organoaluminum cocatalyst, methylalumoxane MAO.

Manufacturing

Melt processing of polypropylene can be achieved via extrusion and molding. Common extrusion methods include production of melt-blown and spun-bond fibres to form long rolls for future conversion into a wide range of useful products, such as face masks, filters, nappies (diapers) and wipes. The most common shaping technique is injection molding, which is used for parts such as cups, cutlery, vials, caps, containers, housewares, and automotive parts such as batteries. The related techniques of blow molding and injection-stretch blow molding are also used, which involve both extrusion and molding.

The large number of end-use applications for polypropylene are often possible because of the ability to tailor grades with specific molecular properties and additives during its manufacture. For example, antistatic additives can be added to help polypropylene surfaces resist dust and dirt. Many physical finishing techniques can also be used on polypropylene, such as machining. Surface treatments can be applied to polypropylene parts in order to promote adhesion of printing ink and paints.

Applications

Since polypropylene is resistant to fatigue, most plastic living hinges, such as those on flip-top bottles, are made from this material. However, it is important to ensure that chain molecules are oriented across the hinge to maximize strength.

Very thin sheets of polypropylene are used as a dielectric within certain high-performance pulse and low-loss RF capacitors.

Polypropylene is used in the manufacturing piping systems; both ones concerned with high-purity and ones designed for strength and rigidity (eg. those intended for use in potable plumbing, hydronic heating and cooling, and reclaimed water). This material is often chosen for its resistance to corrosion and chemical leaching, its resilience against most forms of physical damage, including impact and freezing, its environmental benefits, and its ability to be joined by heat fusion rather than gluing.

Many plastic items for medical or laboratory use can be made from polypropylene because it can withstand the heat in an autoclave. Its heat resistance also enables it to be used as the manufacturing material of consumer-grade kettles. Food containers made from it will not melt in the dishwasher, and do not melt during industrial hot filling processes. For this reason, most plastic tubs for dairy products are polypropylene sealed with aluminum foil (both heat-resistant materials). After the product has cooled, the tubs are often given lids made of a less heat-resistant material, such as LDPE or polystyrene. Such containers provide a good hands-on example of the difference in modulus, since the rubbery (softer, more flexible) feeling of LDPE with respect to polypropylene of the same thickness is readily apparent. Rugged, translucent, reusable plastic containers made in a wide variety of shapes and sizes for consumers from various companies such as Rubbermaid and Sterilite are commonly made of polypropylene, although the lids are often made of somewhat more flexible LDPE so they can snap on to the container to close it. Polypropylene can also be made into disposable bottles to contain liquid, powdered, or similar consumer products, although HDPE and polyethylene terephthalate are commonly also used to make bottles. Plastic pails, car batteries, wastebaskets, pharmacy prescription bottles, cooler containers, dishes and pitchers are often made of polypropylene or HDPE, both of which commonly have rather similar appearance, feel, and properties at ambient temperature.

A common application for polypropylene is as biaxially oriented polypropylene (BOPP). These BOPP sheets are used to make a wide variety of materials including clear bags. When polypropylene is biaxially oriented, it becomes crystal clear and serves as an excellent packaging material for artistic and retail products.

Polypropylene, highly colorfast, is widely used in manufacturing carpets, rugs and mats to be used at home.

Polypropylene is widely used in ropes, distinctive because they are light enough to float in water. For equal mass and construction, polypropylene rope is similar in strength to polyester rope. Polypropylene costs less than most other synthetic fibres.

Polypropylene is also used as an alternative to polyvinyl chloride (PVC) as insulation for electrical cables for LSZH cable in low-ventilation environments, primarily tunnels. This is because it emits less smoke and no toxic halogens, which may lead to production of acid in high-temperature conditions.

Polypropylene is also used in particular roofing membranes as the waterproofing top layer of single-ply systems as opposed to modified-bit systems.

Polypropylene is most commonly used for plastic moldings, wherein it is injected into a mold while molten, forming complex shapes at relatively low cost and high volume; examples include bottle tops, bottles, and fittings. It can also be produced in sheet form, widely used for the production of stationery folders, packaging, and storage boxes.

The wide colour range, durability, low cost, and resistance to dirt make it ideal as a protective cover for papers and other materials. It is used in Rubik's cube stickers because of these characteristics.

The availability of sheet polypropylene has provided an opportunity for the use of the material by designers. The light-weight, durable, and colorful plastic makes an ideal medium for the creation of light shades, and a number of designs have been developed using interlocking sections to create elaborate designs.

Polypropylene sheets are a popular choice for trading card collectors; these come with pockets (nine for standard-size cards) for the cards to be inserted and are used to protect their condition and are meant to be stored in a binder.

Expanded polypropylene (EPP) is a foam form of polypropylene. EPP has very good impact characteristics due to its low stiffness; this allows EPP to resume its shape after impacts. EPP is extensively used in model aircraft and other radio controlled vehicles by hobbyists. This is mainly due to its ability to absorb impacts, making this an ideal material for RC aircraft for beginners and amateurs.

Polypropylene is used in the manufacture of loudspeaker drive units. Its use was pioneered by engineers at the BBC and the patent rights subsequently purchased by Mission Electronics for use in their Mission Freedom Loudspeaker and Mission 737 Renaissance loudspeaker.

Polypropylene fibres are used as a concrete additive to increase strength and reduce cracking and spalling.

Clothing

Polypropylene is a major polymer used in nonwovens, with over 50% used for diapers or sanitary products where it is treated to absorb water (hydrophilic) rather than naturally repelling water (hydrophobic). Other interesting non-woven uses include filters for air, gas, and liquids in which the fibres can be formed into sheets or webs that can be pleated to form cartridges or layers that filter in various efficiencies in the 0.5 to 30 micrometre range. Such applications could be seen in the house as water filters or air-conditioning-type filters. The high surface area and naturally oleophilic polypropylene nonwovens are ideal absorbers of oil spills with the familiar floating barriers near oil spills on rivers.

In New Zealand, in the US military, and elsewhere, polypropylene, or 'polypro' (New Zealand 'polyprops'), has been used for the fabrication of cold-weather base layers, such as long-sleeve shirts or long underwear (More recently, polyester has replaced polypropylene in these applications in the U.S. military, such as in the ECWCS).

Polypropylene is also used in warm-weather gear such as some Under Armour clothing, which can easily transport sweat away from the skin. Although polypropylene clothes are not easily flammable, they can melt, which may result in severe burns if the service member is involved in an explosion or fire of any kind. Polypropylene undergarments are known for retaining body odors which are then difficult to remove. The current generation of polyester does not have this disadvantage.

The material has recently been introduced into the fashion industry through the work of designers such as Anoush Waddington, who have developed specialized techniques to create jewelry and wearable items from polypropylene.

Medical

Its most common medical use is in the synthetic, nonabsorbable suture Prolene, manufactured by Ethicon Inc.

Polypropylene has been used in hernia and pelvic organ prolapse repair operations to protect the body from new hernias in the same location. A small patch of the material is placed over the spot of the hernia, below the skin, and is painless and is rarely, if ever, rejected by the body. However, a polypropylene mesh will erode over the uncertain period from days to years. Therefore, the FDA has issued several warnings on the use of polypropylene mesh medical kits for certain applications in pelvic organ prolapse, specifically when introduced in close proximity to the vaginal wall due to a continued increase in number of mesh erosions reported by patients over the past few years.

EPP Model Aircraft

Since 2001, expanded polypropylene (EPP) foams have been gaining in popularity and in application as a structural material in hobbyist radio control model aircraft. Unlike expanded polystyrene foam (EPS) which is friable and breaks easily on impact, EPP foam is able to absorb kinetic impacts very well without breaking, retains its original shape, and exhibits memory form characteristics which allow it to return to its original shape in a short amount of time. In consequence, a radio-control model whose wings and fuselage are constructed from EPP foam is extremely resilient, and able to absorb impacts that would result in complete destruction of models made from lighter traditional materials, such as balsa or even EPS foams.

EPP models, when covered with inexpensive fibreglass impregnated self adhesive tapes, often exhibit much increased mechanical strength, in conjunction with a lightness and surface finish that rival those of models of the aforementioned types. EPP is also chemically highly inert, permitting the use of a wide variety of different adhesives. EPP can be heat molded, and surfaces can be easily finished with the use of cutting tools and abrasive papers. The principal areas of model making in which EPP has found great acceptance are the fields of:

- Wind-driven Slope Soarers
- Indoor electric powered profile electric models
- Hand launched gliders for small children.

In the field of slope soaring, EPP has found greatest favour and use, as it permits the construction of radio-controlled model gliders of great strength and maneuverability. In consequence, the disciplines of slope combat (the active process of friendly competitors attempting to knock each other's planes out of the air by direct contact) and slope pylon racing have become commonplace, in direct consequence of the strength characteristics of the material EPP.

Repairing

Many objects are made with polypropylene precisely because it is resilient and resistant to most solvents and glues. Also, there are very few glues available specifically for gluing PP. However, solid PP objects not subject to undue flexing can be satisfactorily joined with a two part epoxy glue or using hot-glue guns. Preparation is important and it is often helpful to roughen the surface with a file, emery paper or other abrasive material to provide better anchorage for the glue. Also it is recommended to clean with mineral spirits or similar alcohol prior to gluing to remove any oils or other contamination. Some experimentation may be required. There are also some industrial glues available for PP, but these can be difficult to find, especially in a retail store.

Health Concerns

In 2008, researchers in Canada asserted that quaternary ammonium biocides and oleamide were leaking out of certain polypropylene labware, affecting experimental results. Since polypropylene is used in a wide number of food containers such as those for yogurt, Health Canada media spokesman Paul Duchesne, said the department will be reviewing the findings to determine whether steps are needed to protect consumers.

The Environmental Working Group classifies PP as of low to moderate hazard.

5

Propylene Oxide

Propylene oxide is an organic compound with the molecular formula CH_3CHCH_2O. This colourless volatile liquid is produced on a large scale industrially, its major application being its use for the production of polyether polyols for use in making polyurethane plastics. It is chiral epoxide, although it commonly used as a racemic mixture. This compound is sometimes called 1,2-propylene oxide to distinguish it from its isomer 1,3-propylene oxide, better known as oxetane.

Production

Industrial production of propylene oxide starts from propylene. Two general approaches are employed, one involving hydrochlorination and the other involving oxidation. In 2005, about half of the world production was through chlorohydrin technology and one half via oxidation routes. The latter approach is growing in importance.

Hydrochlorination Route

The traditional route proceeds via the conversion of propylene to chloropropanols:

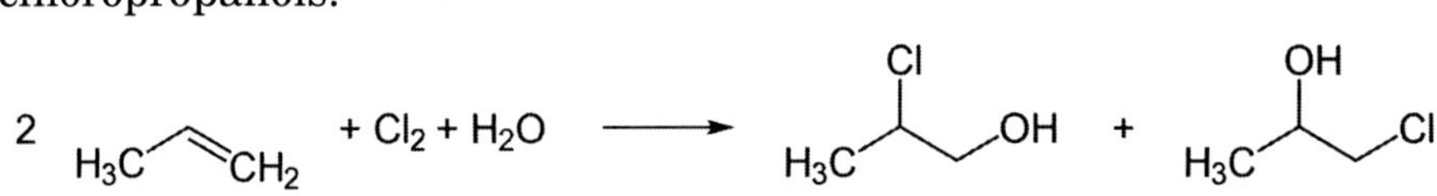

The reaction produces a mixture of 1-chloro-2-propanol and 2-chloro-1-propanol, which is then dehydrochlorinated. For example:

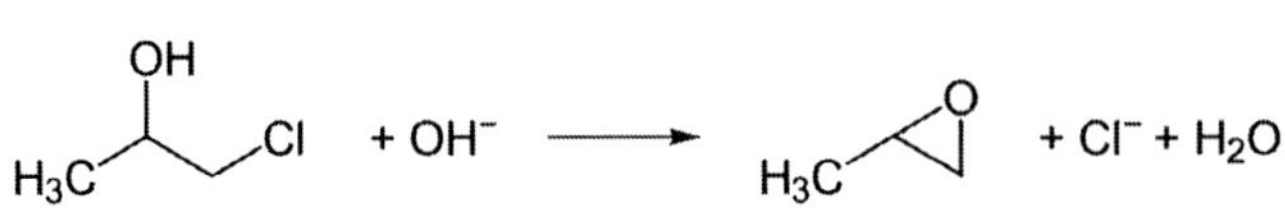

Lime is often used as a chlorine absorber.

Co-oxidation of Propylene

The other general route to propylene oxide involves co-oxidation of the organic chemicals isobutene or ethylbenzene. In the present of catalyst, air oxidation occurs as follows:

$$CH_3CH{=}CH_2 + Ph\text{-}CH_2CH_3 + O_2 \rightarrow CH_3CHCH_2O + Ph\text{-}CH{=}CH_2 + H_2O$$

The coproducts of these reactions, *t*-butyl alcohol or styrene, are useful feedstock for other products. For example, *t*-butyl alcohol reacts with methanol to give MTBE, an additive for gasoline. Before the current ban of MTBE, propylene/isobutene was one of the most important production process.

Oxidation of Propylene

In April 2003, Sumitomo Chemical commercialized the first PO-only plant in Japan, which produces propylene oxide from oxidation of cumene without significant production of other products. This method is a variant of the POSM process (co-oxidation) that uses cumene hydroperoxide instead of ethylbenzene hydroperoxide and recycles the coproduct (alpha-hydroxycumene) via dehydration and hydrogenation back to cumene.

In March 2009, BASF and Dow Chemical started up their new HPPO plant in Antwerp. In the HPPO-Process, propylene is oxidized with hydrogen peroxide:

$$CH_3CH{=}CH_2 + H_2O_2 \rightarrow CH_3CHCH_2O + H_2O$$

In this process no side products other than water are generated.

Uses

Between 60 and 70% of all propylene oxide is converted to polyether polyols for the production of polyurethane plastics. About 20% of propylene oxide is hydrolyzed into propylene glycol, via a process which is accelerated by acid or base catalysis. Other major products are polypropylene glycol, propylene glycol ethers, and propylene carbonate.

Historic and Niche Uses

Propylene oxide was once used as a racing fuel, but that usage is now prohibited under the US NHRA rules for safety reasons. It has also been used in glow fuel for model aircraft and surface vehicles, typically combined in small percentages of around 2% as an additive to the typical methanol, nitromethane, and oil mix. It is also used in thermobaric weapons, and microbial fumigation.

Fumigant

The United States Food and Drug Administration has approved the use of propylene oxide to pasteurize raw almonds beginning on September 1, 2007 in response to two incidents of contamination by *Salmonella* in commercial orchards, one incident occurring in Canada, and one incident in the United States.

Pistachio nuts can also be subjected to propylene oxide to control *Salmonella*. It is a method approved by the FDA.

Microscopy

Propylene oxide is commonly used in the preparation of biological samples for electron microscopy, to remove residual ethanol previously used for dehydration.

In a typical procedure, the sample is first immersed in a mixture of equal volumes of ethanol and propylene oxide for 5 minutes, and then four times in pure oxide, 10 minutes each.

Safety

Propylene oxide is a probable human carcinogen, and listed as an IARC Group 2B carcinogen.

Propylene Glycol

Propylene glycol, also called 1,2-propanediol or propane-1,2-diol, is an organic compound (a diol or double alcohol) with formula $C_3H_8O_2$ or HO-CH_2-CHOH-CH_3. It is a colorless, nearly odorless, clear, viscous liquid with a faintly sweet taste, hygroscopic and miscible with water, acetone, and chloroform.

The compound is sometimes called α-propylene glycol to distinguish it from the isomer propane-1,3-diol HO-$(CH_2)_3$-OH, also called β-propylene glycol.

Chirality

Propylene glycol contains an asymmetrical carbon atom, so it exists in two stereoisomers. The commercial product is a racemic mixture. Pure optical isomers can be obtained by hydration of optically pure propylene oxide.

Production

Industrially, propylene glycol is produced from propylene oxide. Different manufacturers use either non-catalytic high-temperature process at 200 °C (392 °F) to 220 °C (428 °F), or a catalytic method, which proceeds at 150 °C (302 °F) to 180 °C (356 °F) in the presence of ion exchange resin or a small amount of sulfuric acid or alkali.

Final products contain 20% 1,2-propanediol, 1.5% of dipropylene glycol and small amounts of other polypropylene glycols. Further purification produces finished industrial grade or USP/JP/EP/BP grade propylene glycol that is typically 99.5% or greater. Propylene glycol can also be converted from glycerol, a biodiesel byproduct.

Applications

Propylene glycol is used:

- As an ingredient in the oil dispersant Corexit, used in great quantities during the Deepwater Horizon oil spill.
- As a solvent in many pharmaceuticals, including oral, injectable and topical formulations. Notably, diazepam, which is insoluble in water, uses propylene glycol as its solvent in its clinical, injectable form.
- As a humectant food additive, labeled as E number E1520
- As an emulsification agent in Angostura and orange bitters
- As a moisturizer in medicines, cosmetics, food, toothpaste, shampoo, mouth wash, hair care and tobacco products
- As a carrier in fragrance oil
- To produce polyester compounds
- As a base in deicing solution
- As an ingredient in massage oils
- In hand sanitizers, antibacterial lotions, and saline solutions
- In smoke machines to make artificial smoke for use in firefighters' training and theatrical productions
- In electronic cigarettes, as a vaporizable base for diluting the nicotine liquid
- As a solvent for food colours and flavorings
- As an ingredient, along with wax and gelatin, in the production of paintballs

- As a moisture stabilizer (humectant) for snus (Swedish style snuff).
- As a cooling agent for beer and wine glycol jacketed fermentation tanks
- As a non-toxic antifreeze for winterizing drinking water systems, and in applications where the used antifreeze eventually will be drained into the soil, water, or a septic system.
- As a less-toxic antifreeze in solar water heating systems
- As a solvent used in mixing photographic chemicals, such as film developers
- In cryonics
- As a working fluid in hydraulic presses
- As a coolant in liquid cooling systems
- To regulate humidity in a cigar humidor
- As the killing and preserving agent in pitfall traps, usually used to capture ground beetles
- As an additive to pipe tobacco to prevent dehydration.
- To treat livestock ketosis
- As the main ingredient in deodorant sticks.
- To de-ice aircraft.
- As an ingredient in UV or blacklight tattoo ink
- As a lubricant in air conditioning compressors.
- As a wetting agent, used to determine drying time in paints and coatings.

Propylene glycol has properties similar to those of highly toxic ethylene glycol (monoethylene glycol, or MEG). (Note: Infrequently propylene glycol may also use the acronym MEG, but as an abbreviation of methyl ethyl glycol- the industry standard acronym for propylene glycol is PG or MPG (monopropylene glycol)). The industrial norm is to replace ethylene glycol with propylene glycol when safer properties are desired.

Properties

Propylene glycol is a component in newer automotive antifreezes and de-icers used at airports. Like ethylene glycol, the freezing point of propylene glycol is depressed when mixed with water due to disruption of hydrogen bonding. Unlike ethylene glycol, propylene glycol is of very low toxicity. Both are readily biodegradable.

Table 1 : Freezing Points of Propylene Glycol – Water Mixtures

Percent Propylene Glycol (wt. %)	Freezing Point (°F)	Freezing Point (°C)
0	32	0
10	26	-3
20	20	-7
30	10	-12
36	0	-18
40	-5	-20
43	-10	-23
48	-20	-29
52	-30	-34
55	-40	-40
58	-50	-46
60	-60	-51

Safety

Humans

The acute oral toxicity of propylene glycol is very low, and large quantities are required to cause perceptible health damage in humans; propylene glycol is metabolized in the human body into pyruvic acid (a normal part of the glucose-metabolism process, readily converted to energy), acetic acid (handled by ethanol-metabolism), lactic acid (a normal acid generally abundant during digestion), and propionaldehyde (a highly toxic substance). Serious toxicity generally occurs only at plasma concentrations over 1 g/L, which requires extremely high intake over a relatively short period of time. It would be nearly impossible to reach toxic levels by consuming foods or supplements, which contain at most 1 g/kg of PG. Cases of propylene glycol poisoning are usually related to either inappropriate intravenous administration or accidental ingestion of large quantities by children. The potential for long-term oral toxicity is also low. In one study, rats were provided with feed containing as much as 5% PG in feed over a period of 104 weeks and they showed no apparent ill effects. Because of its low chronic oral toxicity, propylene glycol was classified by the U. S. Food and Drug Administration as "generally recognized as safe" (GRAS) for use as a direct food additive.

Prolonged contact with propylene glycol is essentially non-irritating to the skin. Undiluted propylene glycol is minimally irritating to the eye, and can produce slight transient conjunctivitis (the eye recovers after the exposure is removed). Exposure to mists may cause eye irritation, as well as upper respiratory tract irritation. Inhalation of the propylene glycol vapours appears to present no significant hazard in ordinary applications. However, limited human experience indicates that inhalation of propylene glycol mists could be irritating to some individuals. Some research has suggested that propylene glycol not be used in applications where inhalation exposure or human eye contact with the spray mists of these materials is likely, such as fogs for theatrical productions or antifreeze solutions for emergency eye wash stations.

Propylene glycol does not cause sensitization and it shows no evidence of being a carcinogen or of being genotoxic.

Adverse responses to intravenous administration of drugs which use PG as an excipient have been seen in a number of people, particularly with large dosages thereof. Responses may include "hypotension[,] bradycardia... QRS and T abnormalities on the ECG[,] arrhythmia[,] cardiac arrest[,] serum hyperosmolality[,] lactic acidosis[,] and haemolysis".

A high percentage (12% to 42%) of directly-injected propylene glycol is eliminated/secreted in urine unaltered depending on dosage, with the remainder appearing in its glucuronide-form. The speed of renal filtration decreases as dosage increases, which may be due to propylene glycol's mild anesthetic/CNS-depressant -properties as an alcohol. In one case, administration via IV of PG-suspended nitroglycerin to an elderly man may have induced coma and acidosis.

According to a 2010 study by Karlstad University, the concentrations of PGEs, propylene glycol and glycol ethers in indoor air, particularly bedroom air, has been linked to increased risk of developing numerous respiratory and immune disorders in children, including asthma, hay fever, eczema, and allergies, with increased risk ranging from 50% to 180%. This concentration has been linked to use of water-based paints and water-based cleansers.

Animals

Propylene glycol is an approved food additive for dog food under the category of animal feed and is generally recognized as safe (GRAS) for dogs. Similarly, propylene glycol is an approved food additive for human food as well. The exception is that it is prohibited for use in food for cats because of a species-specific reaction in the body, as noted in 21 CFR 582.1666.

Veterinary data indicate that propylene glycol is toxic to dogs with a 50% chance of being lethal at doses of 9mL/kg, although the figure is higher for most laboratory animals (LD_{50} at levels of 20mL/kg).

However, propylene glycol may be toxic to cats in ways not seen in other animals. The U.S. Food and Drug Administration has determined that its presence in or on cat food has not been shown by adequate scientific data to be safe for use. Any such use is considered an adulteration of the cat food and a violation of the Federal Food, Drug, and Cosmetic Act.

Allergic Reaction

Research has suggested that individuals who cannot tolerate propylene glycol probably experience a special form of irritation, but that they only rarely develop allergic contact dermatitis. Other investigators believe that the incidence of allergic contact dermatitis to propylene glycol may be greater than 2% in patients with eczema.

Patients with vulvodynia and interstitial cystitis may be especially sensitive to propylene glycol. Women struggling with yeast infections may also notice that some OTC creams can cause intense burning. Post menopausal women who require the use of an estrogen cream may notice that brand name creams made with propylene glycol often create extreme, uncomfortable burning along the vulva and perianal area. In these cases, patients can request that a local compounding pharmacy make a “propylene glycol free” cream.

Additionally, some electronic cigarette users who inhale propylene glycol vapour may experience dryness of the throat or shortness of breath. As an alternative, some suppliers will put Vegetable Glycerin in the “e-liquid” for those who are allergic (or have bad reactions) to propylene glycol. A Swedish study published in 2010 strongly suggests a connection between airbourne concentrations of propylene glycol in houses and development of asthma and allergic reactions, such as rhinitis, in children.

Possible Air Germicide

Studies conducted in 1942 by Dr. Oswald Hope Robertson of University of Chicago's Billings Hospital showed vaporized propylene glycol inhalation in laboratory mice may prevent pneumonia, influenza, and other respiratory diseases. Additional studies in monkeys and other animals were undertaken to determine long-term effects, especially the potential for accumulation in the lungs. After a few months of treatment, no ill effects were discovered.

Environmental

Propylene glycol is known to exert high levels of biochemical oxygen demand (BOD) during degradation in surface waters. This process can adversely affect aquatic life by consuming oxygen aquatic organisms need to survive. Large quantities of dissolved oxygen (DO) in the water column are consumed when microbial populations decompose ethylene glycol.

The oxygen depletion potential of airport deicing operation discharges is many times greater than that of raw sewage. For example, before application, Type I propylene glycol-based deicing fluid is generally diluted to a mixture containing approximately 50% propylene glycol.

Pure propylene glycol has a five-day biochemical oxygen demand (BOD5) concentration of approximately 1,000,000 mg/L. A typical diluted propylene-based deicing fluid could therefore have a BOD5 concentration of approximately 500,000 mg/L.

In comparison, raw sewage typically has a BOD5 concentration of approximately 200 mg/L. The amount of fluid used to deice a single jet depends on the nature of the precipitation event and the size of the aircraft but can range from several hundred to several thousand gallons.

Therefore, deicing a single jet can generate a BOD5 load greater than that of one million gallons of raw sewage. A large hub airport often has several hundred flights each day.

Sufficient DO levels in surface waters are critical for the survival of fish, macroinvertebrates, and other aquatic organisms. If oxygen concentrations drop below a minimum level, organisms emigrate, if able and possible, to areas with higher oxygen levels or eventually die. This effect can drastically reduce the amount of usable aquatic habitat. Reductions in DO levels can reduce or eliminate bottom-feeder populations, create conditions that favour a change in a community's species profile, or alter critical food-web interactions.

Corexit Implications

The chemical is an ingredient in the oil dispersant Corexit, used in great quantities during the Deepwater Horizon oil spill. Corexit has come under scrutiny for probable adverse effects on marine life and humans that are exposed to it. Propylene glycol has also come under scrutiny, as it is the chemical that disperses Corexit and the oil to subsurface depths.

Acrylic Acid

Acrylic acid (IUPAC: prop-2-enoic acid) is an organic compound with the formula $CH_2{=}CHCO_2H$. It is the simplest unsaturated carboxylic acid, consisting of a vinyl group connected directly to a carboxylic acid terminus. This colorless liquid has a characteristic acrid or tart smell. It is miscible with water, alcohols, ethers, and chloroform. More than one billion kilograms are produced annually.

Production

Acrylic acid is produced from propene which is a byproduct of ethylene and gasoline production.

$$CH_2{=}CHCH_3 + 1.5\ O_2 \rightarrow CH_2{=}CHCO_2H + H_2O$$

Because acrylic acid and its esters have long been valued commercially, many other methods have been developed but most have been abandoned for economic or environmental reasons. An early method was the hydrocarboxylation of acetylene ("Reppe chemistry"):

$$HCCH + CO + H_2O \rightarrow CH_2{=}CHCO_2H$$

This method requires nickel carbonyl and high pressures of carbon monoxide. It was once manufactured by the hydrolysis of acylonitrile which is derived from propene by ammoxidation, but was abandoned because the method cogenerates ammonium derivatives. Other now abandoned precursors to acrylic acid include ethenone and ethylene cyanohydrin.

Reactions and Uses

Acrylic acid undergoes the typical reactions of a carboxylic acid and, when reacted with an alcohol, it will form the corresponding ester. The esters and salts of acrylic acid are collectively known as acrylates (or propenoates). The most common alkyl esters of acrylic acid are methyl-, butyl-, ethyl-, and 2-ethylhexyl-acrylate.

Acrylic acid and its esters readily combine with themselves (to form polyacrylic acid) or other monomers (e.g. acrylamides, acrylonitrile, vinyl, styrene, and butadiene) by reacting at their double bond, forming homopolymers or copolymers which are used in the manufacture of various plastics, coatings, adhesives, elastomers, as well as floor polishes, and paints.

Substituents

As a substituent acrylic acid can be found as an acyl group or a carboxyalkyl group depending on the removal of the group from the molecule. More specifically these are:

1. The acryloyl group, with the removal of the -OH from carbon-1.
2. The 2-carboxyethenyl group, with the removal of a -H from carbon-3. This substituent group is found in chlorophyll.

Safety

Acrylic acid is severely irritating and corrosive to the skin and the respiratory tract. Eye contact can result in severe and irreversible injury. Low exposure will cause minimal or no health effects, while high exposure could result in pulmonary edema. The LD_{50} is 340 mg/kg (rat, oral).

Acrylate Polymer

An acrylate polymer belongs to a group of polymers which could be referred to generally as plastics. They are noted for their transparency and resistance to breakage and elasticity. Also commonly known as acrylics or polyacrylates.

Typical acrylate monomers used to form acrylate polymers are: acrylic acid, methyl methacrylate and acrylonitrile.

Monomers

Other examples of acrylate monomers are:

- Methacrylates
- methyl acrylate
- ethyl acrylate
- 2-chloroethyl vinyl ether
- 2-ethylhexyl acrylate,
- hydroxyethyl methacrylate
- butyl acrylate
- butyl methacrylate
- TMPTA.

Acrylic Elastomers

Acrylic elastomer is a general term for a type of synthetic rubber whose main component is acrylic acid alkylester (ethyl or butyl ester). Acrylic elastomer has characteristics of heat and oil resistance.

It is divided into old type and new type: Old types include ACM (copolymer of acrylic acid ester and 2-chloroethyl vinyl ether) containing chlorine and ANM (copolymer of acrylic acid ester and acrylonitrile) without chloride.

Other than the slightly better water resistance of ANM, there are no physical differences; even processability is poor for both types. Since prices are also high, demand is not so high vis-à-vis the characteristics. On the other hand, the new type of acrylic rubber does not contain any chlorine despite its unclear chemical composition. Processability has been improved, and most of tackiness to rolls as well as staining problems related to molds have been solved.

Major characteristics of acrylic rubber include heat resistance and oil resistance; it can endure a temperature of 170 ~ 180! under dry heat or in oil. Since it does not have a double bond, acrylic rubber also boasts of good weatherability and ozone resistance.

Its cold resistance is not that good, however. The saturation point is -15! for the old type and -28 ~ -30! for the new type. In terms of vulcanization, the standard method for the old type is amine vulcanization. To minimize permanent deformation, the old type requires curing for 24 hours under a temperature of 150!. On the other hand, for the new type, the press curing time and follow-up vulcanization time are significantly reduced by combining metal soap and sulfur. It has no special characteristics. The rebound resilience and abrasion resistance of the new type are poor, and even its electrical characteristics are considerably poor compared with acrylonitrile-butadiene rubber and butyl rubber.

The materials are used mainly for oil seals and packagings related to automobiles.

Other Acrylic Polymers

- Polymethyl methacrylate, an acrylate polymer familiar to consumers is the clear break resistant glass or sheeting sold in hardware stores as acrylic glass or under the trade name Plexiglas.
- Polyacrylate emulsion, water-born coating, are used as binder for outdoor and indoor "latex" house paints
- Acrylic paints as artist paints
- Acrylic fibre
- Sodium polyacrylate water soluble thickeners, a polymer for the production of the Superabsorbent polymer (SAP) used in disposable diapers due to its high absorbency per unit mass
- Acrylic resin as pressure-sensitive adhesive
- "Super glue" is a formulation of cyanoacrylate.
- PVAc copolymer emulsion adhesive of vinyl acetate (VAM) and acrylic acid (VAA)
- Polyacrylamide copolymer used as floculation agent in water treatment.

Allyl Chloride

Allyl chloride is the organic compound with the formula $CH_2=CHCH_2Cl$. This colorless liquid is insoluble in water but soluble in common organic solvents. It is mainly converted to epichlorohydrin, used in the production of plastics. It is a chlorinated derivative of propylene.

Production

Allyl chloride is prepared by the reaction of propylene with chlorine. At lower temperatures, the main product is 1,2-dichloropropane, but at 500 °C, allyl chloride predominates, being formed via a free radical reaction:

$$CH_3CH=CH_2 + Cl_2 \rightarrow ClCH_2CH=CH_2 + HCl$$

An estimated 800M kg are produced in this way annually.

Reactions and Uses

The great majority of allyl chloride is converted to epichlorohydrin. Other commercially significant derivatives include allyl alcohol, allylamine, allyl isothiocyanate (synthetic mustard oil) and allylsilane.

In the laboratory, allyl chloride is a common alkylating agent, useful in the manufacture of pharmaceuticals and pesticides. Illustrative of its reactivity include its cyanation to allyl cyanide ($CH_2{=}CHCH_2CN$), reductive coupling to give diallyl, oxidative addition to palladium to give allylpalladium chloride dimer, $(C_3H_5)_2Pd_2Cl_2$, and dehydrohalogenation to cyclopropene.

Safety

Allyl chloride is highly toxic and flammable.

Epichlorohydrin

Epichlorohydrin (abbreviated ECH) is an organochlorine compound and an epoxide. This is a colorless liquid with a pungent, garlic-like odor, moderately soluble in water, but miscible with most polar organic solvents. Epichlorohydrin is a highly reactive compound and is used in the production of glycerol, plastics, epoxy glues and resins, and elastomers. In contact with water, epichlorohydrin hydrolyzes to 3-MCPD, a carcinogen found in food.

Production

Epichlorohydrin is manufactured from allyl chloride in two steps, beginning with the hydrochlorination using hypochlorous acid, which affords a mixture of two alcohols:

$$CH_2{=}CHCH_2Cl + HOCl \rightarrow HOCH_2CHClCH_2Cl \text{ and, or } ClCH_2CH(OH)CH_2Cl$$

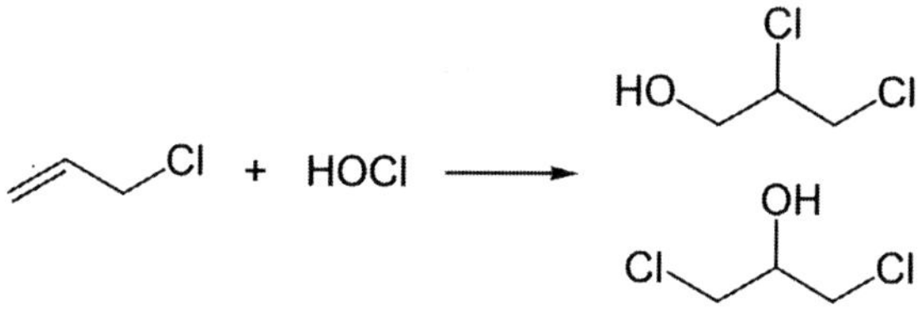

In the second step, this mixture is treated with base to give the epoxide:

$$HOCH_2CHClCH_2Cl \text{ and, or } ClCH_2CH(OH)CH_2Cl + NaOH \rightarrow CH_2CHOCH_2Cl + NaCl + H_2O$$

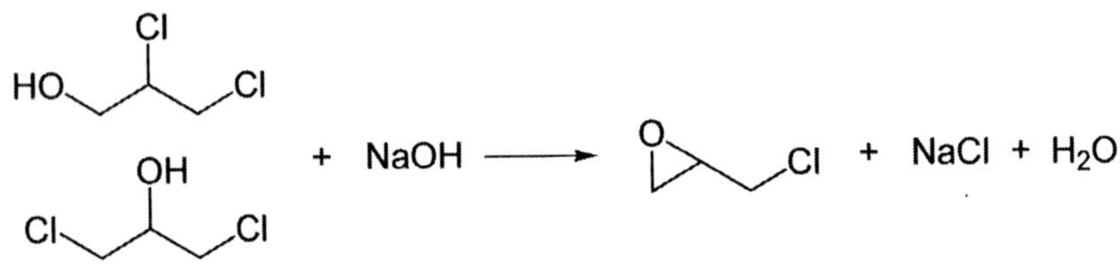

In this way, more than 700,000 tons of epichlorohydrin are produced annually.

Applications

Glycerol and Epoxy Resins Synthesis

Epichlorohydrin is mainly converted to Bisphenol A diglycidyl ether, a building block in the manufacture of epoxy resins. It is also a precursor to monomers for other resins and polymers.

Another usage is the conversion to synthetic glycerol:

$$CH_2CHOCH_2Cl + 2\ H_2O \rightarrow HOCH_2CH(OH)CH_2(OH) + HCl$$

However, the rapid increase in biodiesel production, where Glycerol is a waste product, has led to a glut of glycerol on the market, rendering this process uneconomic for the mass market. Synthetic glycerol is now used only in sensitive pharmaceutical, technical and personal care applications where quality standards are very high. In fact, ways to perform the reverse process are now being researched, to produce epichlorohydrin from the now less valuable natural glycerol, a process now commercialised by Solvay SA with a commercial demo-site producing 10 000 tons annually.

Minor and Niche Applications

Epichlorohydrin is a versatile precursor in the synthesis of many organic compounds. For example, it is converted to glycidyl nitrate, an energetic binder used in explosive and propellant compositions. The epichlorohydrin is reacted with an alkali nitrate, such as sodium nitrate, producing glycidyl nitrate and alkali chloride. It is used as a solvent for cellulose, resins, and paints, and it has found use as an insect fumigant. Epichlorohydrin is used in paper reinforcement (used for instance in the food industry to manufacture tea bags, coffee filters, and sausage/salami casings) and water purification. An important biochemical application of epichlorohydrin is its use as crosslinking agent for the production of Sephadex size-exclusion chromatographic resins from dextrans.

Safety

Epichlorohydrin is irritating and moderately toxic as well as carcinogenic.

Epoxy

Epoxy, also known as polyepoxide, is a thermosetting polymer formed from reaction of an epoxide "resin" with polyamine "hardener". Epoxy has a wide range of applications, including fibre-reinforced plastic materials and general purpose adhesives.

Chemistry

Epoxy is a copolymer; that is, it is formed from two different chemicals. These are referred to as the "resin" or "compound" and the "hardener" or "activator". The resin consists of monomers or short chain polymers with an epoxide group at either end. Most common epoxy resins are produced from a reaction between epichlorohydrin and bisphenol-A, though the latter may be replaced by similar chemicals. The hardener consists of polyamine monomers, for example Triethylenetetramine (TETA). When these compounds are mixed together, the amine groups react with the epoxide groups to form a covalent bond. Each NH group can react with an epoxide group, so that the resulting polymer is heavily crosslinked, and is thus rigid and strong.

The process of polymerization is called "curing", and can be controlled through temperature, choice of resin and hardener compounds, and the ratio of said compounds; the process can take minutes to hours. Some formulations benefit from heating during the cure period, whereas others simply require time and ambient temperatures.

History

The first commercial attempts to prepare resins from epichlorohydrin were made in 1927 in the United States. Credit for the first synthesis of bisphenol-A-based epoxy resins is shared by Dr. Pierre Castan of Switzerland and Dr. S.O. Greenlee of the United

States in 1936. Dr. Castan's work was licensed by Ciba, Ltd. of Switzerland, which went on to become one of the three major epoxy resin producers worldwide. Ciba's epoxy business was spun off and later sold in the late 1990s and is now the advanced materials business unit of Huntsman Corporation of the United States. Dr. Greenlee's work was for the firm of Devoe-Reynolds of the United States. Devoe-Reynolds, which was active in the early days of the epoxy resin industry, was sold to Shell Chemical (now Hexion, formerly Resolution Polymers and others).

Applications

The applications for epoxy-based materials are extensive and include coatings, adhesives and composite materials such as those using carbon fibre and fiberglass reinforcements (although polyester, vinyl ester, and other thermosetting resins are also used for glass-reinforced plastic). The chemistry of epoxies and the range of commercially available variations allows cure polymers to be produced with a very broad range of properties. In general, epoxies are known for their excellent adhesion, chemical and heat resistance, good-to-excellent mechanical properties and very good electrical insulating properties. Many properties of epoxies can be modified (for example silver-filled epoxies with good electrical conductivity are available, although epoxies are typically electrically insulating). Variations offering high thermal insulation, or thermal conductivity combined with high electrical resistance for electronics applications, are available.

Paints and Coatings

Two part epoxy coatings were developed for heavy duty service on metal substrates and use less energy than heat-cured powder coatings. These systems use a 4:1 by volume mixing ratio, and dry quickly providing a tough, protective coating with excellent hardness. Their low volatility and water clean up makes them useful for factory cast iron, cast steel, cast aluminium applications and reduces exposure and flammability issues associated with solvent-borne coatings. They are usually used in industrial and automotive applications since they are more heat resistant than latex-based and alkyd-based paints. Epoxy paints tend to deteriorate, known as chalk out, due to UV exposure.

Polyester epoxies are used as powder coatings for washers, driers and other "white goods". Fusion Bonded Epoxy Powder Coatings (FBE) are extensively used for corrosion protection of steel pipes and fittings used in the oil and gas industry, potable water transmission pipelines (steel), concrete reinforcing rebar, *et cetera*. Epoxy coatings are also widely used as primers to improve the adhesion of automotive and marine paints especially on metal surfaces where corrosion (rusting) resistance is important. Metal cans and containers are often coated with epoxy to prevent rusting, especially for foods like tomatoes that are acidic. Epoxy resins are also used for high performance and decorative flooring applications especially terrazzo flooring, chip flooring and coloured aggregate flooring.

Adhesives

Epoxy adhesives are a major part of the class of adhesives called "structural adhesives" or "engineering adhesives" (that includes polyurethane, acrylic, cyanoacrylate, and other chemistries.) These high-performance adhesives are used in the construction of aircraft, automobiles, bicycles, boats, golf clubs, skis, snowboards, and other applications where high strength bonds are required. Epoxy adhesives can be developed to suit almost any application. They can be used as adhesives for wood, metal, glass, stone, and some plastics. They can be made flexible or rigid, transparent or opaque/coloured, fast setting or slow setting. Epoxy adhesives are better in heat and chemical resistance than other common adhesives. In general, epoxy adhesives cured with heat will be more heat- and chemical-resistant than those cured at room temperature. The strength of epoxy adhesives is degraded at temperatures above 350 °F (177 °C).

Some epoxies are cured by exposure to ultraviolet light. Such epoxies are commonly used in optics, fibre optics, optoelectronics, and dentistry.

Industrial Tooling and Composites

Epoxy systems are used in industrial tooling applications to produce molds, master models, laminates, castings, fixtures, and other industrial production aids. This "plastic tooling" replaces metal, wood and other traditional materials, and generally improves the efficiency and either lowers the overall cost or shortens the lead-time for many industrial processes. Epoxies are also used in producing fibre-reinforced or composite parts. They are more expensive than polyester resins and vinyl ester resins, but usually produce stronger and more temperature-resistant composite parts.

Electrical Systems and Electronics

Epoxy resin formulations are important in the electronics industry, and are employed in motors, generators, transformers, switchgear, bushings, and insulators. Epoxy resins are excellent electrical insulators and protect electrical components from short circuiting, dust and moisture. In the electronics industry epoxy resins are the primary resin used in overmolding integrated circuits, transistors and hybrid circuits, and making printed circuit boards. The largest volume type of circuit board—an "FR-4 board"—is a sandwich of layers of glass cloth bonded into a composite by an epoxy resin. Epoxy resins are used to bond copper foil to circuit board substrates, and are a component of the solder mask on many circuit boards.

Flexible epoxy resins are used for potting transformers and inductors. By using vacuum impregnation on uncured epoxy, winding-to-winding, winding-to-core, and winding-to-insulator air voids are eliminated. The cured epoxy is an electrical insulator and a much better conductor of heat than air. Transformer and inductor hot spots are greatly reduced, giving the component a stable and longer life than unpotted product.

Epoxy resins are applied using the technology of resin dispensing.

Consumer and Marine Applications

Epoxies are sold in hardware stores, typically as a pack containing separate resin and hardener, which must be mixed immediately before use. They are also sold in boat shops as repair resins for marine applications. Epoxies typically are not used in the outer layer of a boat because they deteriorate by exposure to UV light. They are often used during boat repair and assembly, and then over-coated with conventional or two-part polyurethane paint or marine-varnishes that provide UV protection.

There are two main areas of marine use. Because of the better mechanical properties relative to the more common polyester resins, epoxies are used for commercial manufacture of components where a high strength/weight ratio is required. The second area is that their strength, gap filling properties and excellent adhesion to many materials including timber have created a boom in amateur building projects including aircraft and boats.

Normal gelcoat formulated for use with polyester resins and vinylester resins does not adhere to epoxy surfaces, though epoxy adheres very well if applied to polyester resin surfaces. “Flocoat” that is normally used to coat the interior of polyester fibreglass yachts is also compatible with epoxies.

Epoxy materials tend to harden somewhat more gradually, while polyester materials tend to harden quickly, particularly if a lot of catalyst is used. The chemical reactions in both cases are exothermic. Large quantities of mix will generate their own heat and greatly speed the reaction, so it is usual to mix small amounts which can be used quickly.

While it is common to associate polyester resins and epoxy resins, their properties are sufficiently different that they are properly treated as distinct materials. Polyester resins are typically low strength unless used with a reinforcing material like glass fibre, are relatively brittle unless reinforced, and have low adhesion. Epoxies, by contrast, are inherently strong, somewhat flexible and have excellent adhesion. However, polyester resins are much cheaper.

Epoxy resins typically require a precise mix of two components which form a third chemical. Depending on the properties required, the ratio may be anything from 1:1 or over 10:1, but in every case they must be mixed exactly. The final product is then a precise thermo-setting plastic. Until they are mixed the two elements are relatively inert, although the ‘hardeners’ tend to be more chemically active and should be protected from the atmosphere and moisture. The rate of the reaction can be changed by using different hardeners, which may change the nature of the final product, or by controlling the temperature.

By contrast, polyester resins are usually made available in a ‘promoted’ form, such that the progress of previously-mixed resins from liquid to solid is already underway, albeit very slowly. The only variable available to the user is to change the rate of this process using a catalyst, often Methyl-Ethyl-Ketone-Peroxide (MEKP), which is very toxic. The presence of the catalyst in the final product actually detracts from the desirable properties, so that small amounts of catalyst are preferable, so long as the hardening proceeds at an acceptable pace. The rate of cure of polyesters can therefore be controlled by the amount and type of catalyst as well as by the temperature. As adhesives, epoxies bond in three ways: a) Mechanically, because the bonding surfaces are roughened; b) By proximity, because the cured resins are physically so close to the bonding surfaces that they are hard to separate; c) Ionically, because the epoxy resins form ionic bonds at an atomic level with the bonding surfaces. This last is substantially the strongest of the three. By contrast, polyester resins can only bond using the first two of these, which greatly reduces their utility as adhesives and in marine repair.

Aerospace Applications

In the aerospace industry, epoxy is used as a structural matrix material which is then reinforced by fibre. Typical fibre reinforcements include glass, carbon, Kevlar, and boron. Epoxies are also used as a structural glue. Materials like wood, and others that are 'low-tech' are glued with epoxy resin.

Art

Epoxy resin, mixed with pigment, may be used as a painting medium, by pouring layers on top of each other to form a complete picture.

Industry

As of 2006, the epoxy industry amounts to more than US$5 billion in North America and about US$15 billion worldwide. The Chinese market has been growing rapidly, and accounts for more than 30% of the total worldwide market. It is made up of approximately 50–100 manufacturers of basic or commodity epoxy resins and hardeners.

These commodity epoxy manufacturers mentioned above typically do not sell epoxy resins in a form usable to smaller end users, so there is another group of companies that purchase epoxy raw materials from the major producers and then compounds (blends, modifies, or otherwise customizes) epoxy systems from these raw materials. These companies are known as "formulators".

The majority of the epoxy systems sold are produced by these formulators and they comprise over 60% of the dollar value of the epoxy market. There are hundreds of ways that these formulators can modify epoxies—by adding mineral fillers (talc, silica, alumina, etc.), by adding flexibilizers, viscosity reducers, colorants, thickeners, accelerators, adhesion promoters, etc.. These modifications are made to reduce costs, to improve performance, and to improve processing convenience. As a result a typical formulator sells dozens or even thousands of formulations—each tailored to the requirements of a particular application or market.

Impacted by the global economic slump, the epoxy market size declined to $15.8 billion in 2009, almost to the level of 2005. In some regional markets it even decreased nearly 20%. The current epoxy market is experiencing positive growth as the global economy revives. With an annual growth rate of 3.5 - 4% the epoxy market is expected to reach $17.7 billion by 2012 and $21.35 by 2015. Higher growth rate is foreseen thereafter due to stronger demands from epoxy composite market and epoxy adhesive market.

Health Risks

The primary risk associated with epoxy use is sensitization to the hardener, which, over time, can induce an allergic reaction. Allergic reaction sometimes occurs at a time which is delayed several days from the exposure. Epoxy use is a main source of occupational asthma among users of plastics. Bisphenol A, which is used in epoxy resin, is a known endocrine disruptor.

Butene

Butene, also known as butylene, is an alkene with the formula C_4H_8. It is a colourless gas that is present in crude oil as a minor constituent in quantities that are too small for viable extraction. It is therefore obtained by catalytic cracking of long chain hydrocarbons left during refining of crude oil. Cracking produces a mixture of products and the butene is extracted from this by fractional distillation.

Butene can be used as the monomer for polybutene but this polymer is more expensive than alternatives with shorter carbon chains such as polypropylene. Polybutene is therefore commonly used as a co-polymer (mixed with another polymer, either during or after reaction), such as in hot-melt adhesives.

Isomers

Among the molecules which have the chemical formula C_4H_8 four isomers are alkenes. All four of these hydrocarbons have four carbon atoms and one double bond in their molecules, but have different chemical structures. The IUPAC and common names, respectively.

In the chemical structures above, the small blue numbers in the structure images are the numbering of the atoms in the main backbone chain of the molecules. Other organic compounds have the formula C_4H_8, namely cyclobutane and methylcyclopropane, but are not alkenes and are not discussed here. There are also cyclic alkenes with four carbon atoms overall such as cyclobutene and two isomers of methylcyclopropene, but they do not have the formula C_4H_8 and are not discussed here. All four of these isomers are gases at room temperature and pressure, but can be liquefied by lowering the temperature or raising the pressure on them, in a manner similar to pressurised butane. These gases are colourless, but do have distinct odours, and are highly flammable. Although not naturally present in petroleum in high percentages, they can be produced from petrochemicals or by catalytic cracking of petroleum. Although they are stable compounds, the carbon-carbon double bonds make them more reactive than similar alkanes, which are more inert compounds in various ways.

Because of the double bonds, these 4-carbon alkenes can act as monomers in the formation of polymers, as well as having other uses as petrochemical intermediates. They are used in the production of synthetic rubber. But-1-ene is a linear or normal alpha-olefin and isobutylene is a branched alpha-olefin. In a rather low percentage, but-1-ene is used as one of the comonomers, along with other alpha-olefins, in the production of high density polyethylene and linear low density polyethylene. Butyl rubber is made by cationic polymerisation of isobutylene with about 2 - 7% isoprene. Isobutylene is also used for the production of methyl *tert*-butyl ether (MTBE) and isooctane, both of which improve the combustion of gasoline.

Isobutylene

Isobutylene (or 2-methylpropene) is a hydrocarbon of significant industrial importance. It is a four-carbon branched alkene (olefin), one of the four isomers of butylene. At standard temperature and pressure it is a colorless flammable gas.

Uses

Isobutylene is used as an intermediate in the production of a variety of products. It is reacted with methanol and ethanol in the manufacture of the gasoline oxygenates methyl tert-butyl ether (MTBE) and ethyl tert-butyl ether (ETBE), respectively. Alkylation with butane produces isooctane, another fuel additive. Isobutylene is also used in the production of methacrolein. Polymerization of isobutylene produces butyl rubber (polyisobutylene). Antioxidants such as butylated hydroxytoluene (BHT) and butylated hydroxyanisole (BHA) are produced by Friedel-Crafts alkylation of phenols using isobutylene.

Manufacture

Isobutylene can be isolated from refinery streams by reaction with sulfuric acid, but the most common industrial method for its production is by catalytic dehydrogenation of isobutane. In the 1990s, the production of isobutylene increased dramatically as the demand for oxygenates such as MTBE grew. Key manufacturers of this product are Texas Petrochemicals(TPC Group) and Lyondell in North America.

Safety

Isobutylene is a highly flammable gas and presents an explosion danger. Usually stored as a compressed gas, if released it may produce an oxygen-deficient atmosphere that presents an asphyxiation hazard.

Methyl Tert-butyl Ether

Methyl *tert*-butyl ether, also known as methyl tertiary butyl ether and MTBE, is an organic compound with molecular formula $(CH_3)_3COCH_3$. MTBE is a volatile, flammable, and colorless liquid that is immiscible with water.

It has a minty odor vaguely reminiscent of diethyl ether, leading to unpleasant taste and odor in water.

MTBE is a gasoline additive, used as an oxygenate to raise the octane number. Its use is controversial in the US and declining in that country in part because of its occurrence in groundwater and legislation favoring ethanol. Worldwide production of MTBE has been constant at about 18 million tons/y (2005) owing to growth in Asian markets which are less subject to ethanol subsidies.

Production and Properties

MTBE is manufactured via the chemical reaction of methanol and isobutylene. Methanol is derived from natural gas, and isobutylene is derived from butane obtained from crude oil or natural gas, thus MTBE is derived from fossil fuels. In the United States, it was produced in very large quantities (more than 200,000 barrels (32,000 m^3) per day in 1999) during its use as a fuel additive.

Uses

MTBE is almost exclusively used as a fuel component in fuel for gasoline engines. It is one of a group of chemicals commonly known as oxygenates because they raise the oxygen content of gasoline.

As Anti-knocking Agent

In the US it has been used in gasoline at low levels since 1979 to replace tetraethyl lead and to increase its octane rating helping prevent engine knocking. Oxygenates help gasoline burn more completely, reducing tailpipe emissions from pre-1984 motor vehicles; dilutes or displaces gasoline components such as aromatics (e.g., benzene) and sulfur; and optimizes the oxidation during combustion. Most refiners chose MTBE over other oxygenates primarily for its blending characteristics and low cost.

Alternatives to MTBE as an Anti-knock Agent

Other compounds are available as additives for gasoline including ethanol and some ethers such as tert-amyl methyl ether (TAME).

Ethanol has been advertised as a safe alternative by the agricultural interest groups in the USA and Europe. Its lack of toxicity is not different from MTBE, but as a polar solvent, it drives off nonpolar hydrocarbons from the gasoline, a problem that MTBE does not cause. Some volatile hydrocarbons in gasoline vapours are carcinogens, and gasoline vapours produce photochemical smog. Ethanol's higher cost requires government intervention in the form of subsidies or mandated usage to be competitive. In 2003, California was the first U.S. state to start replacing MTBE with ethanol. Several other states started switching soon thereafter.

Advocates of both sides of the debate in the United States sometimes claim that gasoline manufacturers have been forced to add MTBE to gasoline by law. It might be more correct to say they have been induced to do so, although any oxygenate would fulfil the law.

An alternative to straight ethanol is the related ether ETBE, which is manufactured from ethanol and isobutene. Its performance as an additive is similar to MTBE, but due to the higher price of ethanol compared to methanol, it is more expensive.

Higher quality gasoline is also an alternative, i.e., so that additives such as MTBE are unnecessary. Iso-octane itself is used. MTBE plants can be retrofitted to produce iso-octane from isobutylene.

As a Solvent

As a solvent, MTBE possesses one distinct advantage over most ethers - it has a reduced tendency to form explosive organic peroxides. Opened bottles of diethyl ether or THF can build up dangerous levels of these peroxides in months, whereas samples of MTBE are usually safe for years (but they should still be tested periodically). For this reason (as well as its higher boiling point), MTBE is used as a solvent extensively in industry, where safety concerns and regulations make working with diethyl ether, THF, or other ethers much more difficult and expensive. However, despite the popularity of MTBE in industrial settings, it is rarely used as a solvent in academia with some exceptions.

MTBE forms azeotropes with water (52.6 °C; 96.5% MTBE). and methanol (51.3 °C; 68.6% MTBE).

Although an ether, MTBE is a poor Lewis base and does not support formation of Grignard reagents. It is also unstable toward strong acids. It reacts dangerously with bromine.

Persistence and Pervasiveness in the Environment

MTBE gives water an unpleasant taste at very low concentrations, and thus can render large quantities of groundwater non-potable. MTBE is often introduced into water-supply aquifers by leaking underground storage tanks (USTs) at gasoline stations or by gasoline containing MTBE spilled onto the ground.

The higher water solubility and persistence of MTBE cause it to travel faster and farther than many other components of gasoline when released into an aquifer.

MTBE is biodegraded by the action of bacteria. In the proper type of bioreactor, such as a fluidized bed bioreactor, MTBE can be rapidly and economically removed from water to undetectable levels. Activated carbon produced from coconut shells and optimized for MTBE adsorption can also reduce MTBE to undetectable levels.

Health Risks

According to the IARC, a cancer research agency of the World Health Organization, MTBE is not classified as a human carcinogen. MTBE can be tasted in water at concentrations of 5 – 15 μg/l.

As of 2007, researchers have limited data about the health effects of ingestion of MTBE. The United States Environmental Protection Agency (EPA) has concluded that available data are inadequate to quantify health risks of MTBE at low exposure levels in drinking water, but that the data support the conclusion that MTBE is a potential human carcinogen at high doses.

Legislation and Litigation in the U.S.

MTBE removal from groundwater and soil contamination in the U.S. is estimated to cost from $1 billion to $30 billion, including removing the compound from aquifers and municipal water supplies and replacing leaky underground oil tanks. In one case, the cost to oil companies to clean up the MTBE in wells belonging to Santa Monica is estimated to exceed $200 million. In another case, the City of New York estimated a $250 million cost for cleanup of a single wellfield in Queens, NY.

MTBE is banned in the US states of California and New York. MTBE consumption, banned the chemical starting January 1, 2004, and as of September 2005, twenty-five states had signed legislation banning MTBE. (A table of state by state information, as of 2002, is available at the United States Department of Energy website.)

In 2000, the U.S. EPA drafted plans to phase out the use of MTBE nationwide over four years. As of fall 2006, hundreds of lawsuits are still pending regarding MTBE contamination of public and private drinking water supplies.

The Energy Policy Act of 2005, passed in the U.S. House of Representatives, did not include a provision for shielding MTBE manufacturers from water contamination lawsuits. This provision was first proposed in 2003 and had been thought by some to be a priority of Tom DeLay and Rep. Joe Barton, then chairman of the Energy and Commerce Committee.

This bill did include a provision that gives MTBE makers, including some major oil companies, $2 billion in transition assistance as MTBE is phased out over the next nine years. Due to opposition in the Senate, the conference report dropped all MTBE provisions. The final bill was passed by both houses and signed into law by President Bush. The lack of MTBE liability protection is resulting in a switchover to the use of ethanol as a gasoline additive. Some traders and consumer advocates are blaming this for an increase in gasoline prices. The U.S. EPA currently lists MTBE as a candidate for a maximum contaminant level (MCL) in drinking water. MCLs are determined by the EPA using toxicity data.

Isoprene

Isoprene (short for isoterpene), or 2-methyl-1,3-butadiene, is a common organic compound with the formula $CH_2=C(CH_3)CH=CH_2$. Under standard conditions it is a colorless liquid. However, this compound is highly volatile because of its low boiling point.

Isoprene (C_5H_8) is the monomer of natural rubber and also a common structure motif to an immense variety of other naturally occurring compounds, collectively termed the isoprenoids. Molecular formula of isoprenoids are multiples of isoprene in the form of $(C_5H_8)_n$, and this is termed the isoprene rule. The functional isoprene units in biological systems are dimethylallyl diphosphate (DMADP) and its isomer isopentenyl diphosphate (IDP).

The singular terms "isoprene" and "terpene" are synonymous whereas the plurals "isoprenes" or "terpenes" refer to terpenoids (isoprenoids).

Natural Occurrences

Isoprene is produced and emitted by many species of trees into the atmosphere (major producers are oaks, poplars, eucalyptus, and some legumes). The yearly production of isoprene emissions by vegetation is around 600 Tg, with half that coming from tropical broadleaf trees and the remainder coming from shrubs.

This is about equivalent to methane emission into the atmosphere and accounts for ~1/3 of all hydrocarbons released into the atmosphere. After release, isoprene is converted by free radicals (like the hydroxyl (OH) radical) and to a lesser extent by ozone into various species, such as aldehydes, hydroperoxides, organic nitrates, and epoxides, which mix into water droplets and help create aerosols and haze. While most in the field acknowledges that isoprene emission effect aerosol formation, whether isoprene increases or decreases aerosol formation is debated.

A second major effect of isoprene on the atmosphere is that in presence of nitric oxides (NO_x) it contributes to the formation of tropospheric (lower atmosphere) ozone, which is one of the leading air pollutants in many countries.

Isoprene itself is normally not regarded as a pollutant, as it is one of the natural products from plants. Formation of tropospheric ozone is only possible in presence of high levels of NO_x, which comes almost exclusively from industrial activities. In fact, isoprene can have the opposite effect and quench ozone formation under low levels of NO_x.

Isoprene Production from Plants

Isoprene is made through the methyl-erythritol 4-phosphate pathway (MEP pathway, also called the non-mevalonate pathway) in the chloroplasts of plants. One of the two end products of MEP pathway, dimethylallyl diphosphate (DMADP), is catalysed by the enzyme isoprene synthase to form isoprene. Therefore, inhibitors that block the MEP pathway, such as fosmidomycin, also blocks isoprene formation. Isoprene emission increases dramatically with temperature and maximizes at around 40 °C. This has led to the hypothesis that isoprene may protect plants against heat stress. Emission of isoprene is also observed in some bacteria and this is thought to come from non-enzymatic degradations from DMADP.

Regulation of Isoprene Emission

Isoprene emission in plants is controlled both by the availability of substrate (DMADP) and by enzyme (isoprene synthase) activity. In particular, light, CO_2 and O_2 dependencies of isoprene emission are controlled by substrate availability, whereas temperature dependency of isoprene emission is regulated both by substrate level and enzyme activity.

Biological Roles

Isoprene emission appears to be a mechanism that trees use to combat abiotic stresses. In particular, isoprene has been shown to protect against moderate heat stresses (~ 40 °C). It was proposed that isoprene emission was specifically used by plants to protect against large fluctations in leaf temperature.

Isoprene is incorporated into and helps stabilize cell membranes in response to heat stress, conferring some tolerance to heat spikes. Isoprene may also confer some resistance to reactive oxygen species. The amount of isoprene released from isoprene-emitting vegetation depends on leaf mass, leaf area, light (particularly photosynthetic photon flux density, or PPFD), and leaf temperature. Thus, during the night, little isoprene is emitted from tree leaves, whereas daytime emissions are expected to be substantial during hot and sunny days, up to 25 μg/(g dry-leaf-weight)/hour in many oak species.

Isoprene in other Organisms

In general, it is the most common hydrocarbon found in the human body,. The estimated production rate of isoprene in the human body is 0.15 μmol/(kg·h), equivalent to approximately 17 mg/day for a person weighing 70 kg. Isoprene is also common in low concentrations in many foods.

Industrial Production

Isoprene was first isolated by thermal decomposition of natural rubber. It is most readily available industrially as a byproduct of the thermal cracking of naphtha or oil, as a side product in the production of ethylene. About 800,000 tonnes are produced annually. About 95% of isoprene production is used to produce cis-1,4-polyisoprene—a synthetic version of natural rubber.

Natural rubber is a polymer of isoprene—most often cis-1,4-polyisoprene—with a molecular weight of 100,000 to 1,000,000. Typically, a few percent of other materials, such as proteins, fatty acids, resins, and inorganic materials are found in high-quality natural rubber. Some natural rubber sources called gutta percha are composed of trans-1,4-polyisoprene, a structural isomer that has similar, but not identical, properties.

Isoprene as a Structural Motif

Isoprene is a common structural motif in biological systems. The isoprenoids (for example, the carotenes are tetraterpenes) are derived from isoprene. Also derived from isoprene are phytol, retinol (vitamin A), tocopherol (vitamin E), dolichols, and squalene. Heme A has an isoprenoid tail, and lanosterol, the sterol precursor in animals, is derived from squalene and hence from isoprene. The functional isoprene units in biological systems are dimethylallyl diphosphate (DMADP) and its isomer isopentenyl diphosphate (IDP), which are used in the biosynthesis of naturally occurring isoprenoids such as carotenoids, quinones, lanosterol derivatives (e.g. steroids) and the prenyl chains of certain compounds (e.g. phytol chain of chlorophyll).

Amine

Amines are organic compounds and functional groups that contain a basic nitrogen atom with a lone pair. Amines are derivatives of ammonia, wherein one or more hydrogen atoms have been replaced by a substituent such as an alkyl or aryl group. Important amines include amino acids, biogenic amines, trimethylamine, and aniline. Inorganic derivatives of ammonia are also called amines, such as chloramine ($NClH_2$).

Primary amine *Secondary amine* *Tertiary amine*

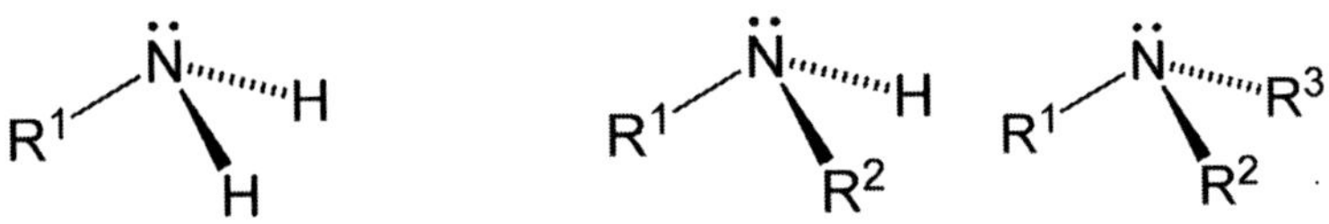

Compounds with the nitrogen atom attached to a carbonyl of the structure R-C(=O)NR'R" are called amides and have different chemical properties from amines.

Classes of Amines

Aliphatic Amines

Primary amines arise when one of three hydrogen atoms in ammonia is replaced by an alkyl. Important primary alkyl amines include methylamine, ethanolamine (2-aminoethanol), and the buffering agent tris. Secondary amines have two alkyl substituents bound to N together with one hydrogen. Important representatives include dimethylamine and methylethanolamine. In tertiary amines, all three hydrogen atoms are replaced by organic substituents. Examples include trimethylamine, a distinctively fishy smell. Cyclic amines are either secondary or tertiary amines. Examples of cyclic amines include the 3-member ring aziridine and the six-membered ring piperidine. N-methylpiperidine is a cyclic tertiary amine. It is also possible to have four alkyl substituents on the nitrogen. These compounds are not amines but are called quaternary ammonium cations, have a charged nitrogen centre, and necessarily come with an anion.

Aromatic Amines

Aromatic amines have the nitrogen atom connected to an aromatic ring as in anilines. The aromatic ring decreases the alkalinity of the amine, depending on its substituents. The presence of an amine group strongly increases the reactivity of the aromatic ring, due to an electron-donating effect.

Naming Conventions

Amines are named in several ways. Typically, the compound is given the prefix "amino-" or the suffix: "-amine." The prefix "N-" shows substitution on the nitrogen atom. An organic compound with multiple amino groups is called a diamine, triamine, tetraamine and so forth.

Systematic names for some common amines:

Lower amines are named with the suffix -amine.

Higher amines have the prefix amino as a functional group.

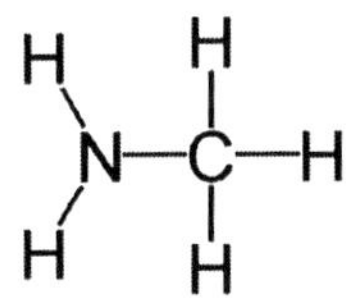

methylamine

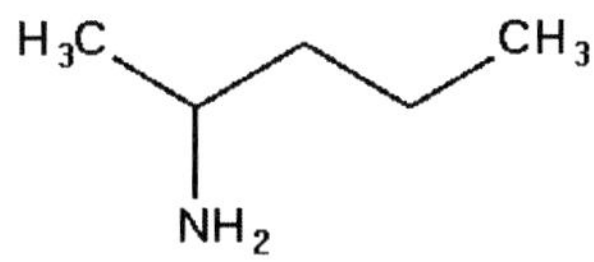

2-aminopentane(or sometimes: pent-2-yl-amine or pentane-2-amine)

Physical Properties

Hydrogen bonding significantly influences the properties of primary and secondary amines. Thus the boiling point of amines is higher than those of the corresponding phosphines, but generally lower than those of the corresponding alcohols. For example, methylamine and ethylamine are gases under standard conditions, whereas the corresponding methyl alcohol and ethyl alcohols are liquids. Gaseous amines possess a characteristic ammonia smell, liquid amines have a distinctive "fishy" smell.

Also reflecting their ability to form hydrogen bonds, most aliphatic amines display some solubility in water. Solubility decreases with the increase in the number of carbon atoms. Aliphatic amines display significant solubility in organic solvents, especially polar organic solvents. Primary amines react with ketones such as acetone.

The aromatic amines, such as aniline, have their lone pair electrons conjugated into the benzene ring, thus their tendency to engage in hydrogen bonding is diminished. Their boiling points are high and their solubility in water low

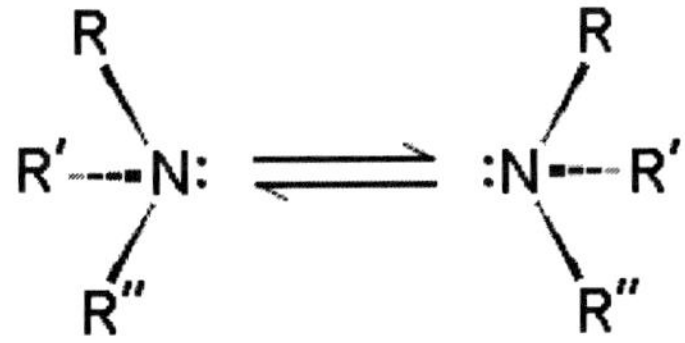

Chirality

Amines of the type NHRR' and NRR'R" are chiral: the nitrogen atom bears four substituents counting the lone pair. The energy barrier for the inversion of the stereocenter is relatively low, e.g., ~7 kcal/mol for a trialkylamine. The interconversion of the stereoisomers has been compared to the inversion of an open umbrella in to a strong wind. Because of this low barrier, amines such as NHRR' cannot be resolved optically and NRR'R" can only be resolved when the R, R', and R" groups are constrained in cyclic structures such as aziridines. Quaternary ammonium salts with four distinct groups on the nitrogen are capable of exhibiting optical activity.

Properties as Bases

Like ammonia, amines are bases. Compared to alkali metal hydroxides, amines are weaker. The basicity of amines depends on:

1. The electronic properties of the substituents (alkyl groups enhance the basicity, aryl groups diminish it).
2. Steric hindrance offered by the groups on nitrogen.
3. The degree of solvation of the protonated amine.

The nitrogen atom features a lone electron pair that can bind H^+ to form an ammonium ion R_3NH^+. The lone electron pair is represented in this article by a two dots above or next to the N. The water solubility of simple amines is largely due to hydrogen bonding between protons in the water molecules and these lone electron pairs.

Table 2 : Inductive effect of alkyl groups

Ions of compound	K_b
Ammonia NH_3	$1.8 \cdot 10^{-5}$ M
Propylamine $CH_3CH_2CH_2NH_2$	$4.7 \cdot 10^{-4}$ M
2-Propylamine $(CH_3)_2CHNH_2$	$3.4 \cdot 10^{-4}$ M
Methylamine CH_3NH_2	$4.4 \cdot 10^{-4}$ M
Dimethylamine $(CH_3)_2NH$	$5.4 \cdot 10^{-4}$ M
Trimethylamine $(CH_3)_3N$	$5.9 \cdot 10^{-5}$ M

+I effect of alkyl groups raises the energy of the lone pair of electrons, thus elevating the basicity. Thus the basicity of an amine may be expected to increase with the number of alkyl groups on the amine. However, there is no strict trend in this regard, as basicity is also governed by other factors mentioned above. Consider the Kb values of the methyl amines given above. The increase in Kb from methylamine to dimethylamine may be attributed to +I effect; however, there is a decrease from dimethylamine to trimethyl amine due to the predominance of steric hindrance offered by the three methyl groups to the approaching Brönsted acid.

Table 3 : Mesomeric effect of aromatic systems

Ions of compound	K_b
Ammonia NH_3	$1.8 \cdot 10^{-5}$ M
Aniline $C_6H_5NH_2$	$3.8 \cdot 10^{-10}$ M
4-Methylaniline 4-$CH_3C_6H_4NH_2$	$1.2 \cdot 10^{-9}$ M
2-Nitroaniline	$1.5 \cdot 10^{-15}$ M
3-Nitroaniline	$2.8 \cdot 10^{-13}$ M
4-Nitroaniline	$9.5 \cdot 10^{-14}$ M

-M effect of aromatic ring delocalises the lone pair of electrons on nitrogen into the ring, resulting in decreased basicity. Substituents on the aromatic ring, and their positions relative to the amine group may also considerably alter basicity as seen above.

The of solvation of protonated amines changes upon their conversion to ammonium compounds. Typically salts of ammonium compounds exhibit the following order of solubility in water: primary ammonium (RNH_3^+)> secondary ammonium ($R_2NH_2^+$). tertiary ammonium (R_3NH^+). Quaternary ammonium salts usually exhibit the lowest solubility of the series.

In sterically hindered amines, as in the case of trimethylamine, the protonated form is not well-solvated. For this reason the parent amine is less basic than expected. In the case of aprotic polar solvents (like DMSO and DMF), wherein the extent of solvation is not as high as in protic polar solvents (like water and methanol), the basicity of amines is almost solely governed by the electronic factors within the molecule.

Synthesis

Alkylation

The most industrially significant amines are prepared from ammonia by alkylation with alcohols:

$$ROH + NH_3 \rightarrow RNH_2 + H_2O$$

These reactions require catalysts, specialized apparatus, and additional purification measures since the selectivity can be problematic. The same amines can be prepared by treatment of Haloalkanes with ammonia and amines:

$$RX + 2\ R'NH_2 \rightarrow RR'NH + [RR'NH_2]X$$

Such reactions, which are most useful for alkyl iodides and bromides, are rarely employed because the degree of alkylation is difficult to control.

Specialized Methods

Many laboratory methods exist for the preparation of amines, many of these methods being rather specialized.

Reaction name	Substrate	Comment
Gabriel synthesis	organohalide	reagent: potassium phthalimide
Staudinger reduction	Azide	This reaction also takes place with a reducing agent such as lithium aluminium hydride.
Schmidt reaction	carboxylic acid	
Aza-Baylis–Hillman reaction	imine	Synthesis of allylic amines
Hofmann degradation	amide	This reaction is valid for preparation of primary amines only. Gives good yields of primary amines uncontaminated with other amines.
Hofmann Elimination	Quaternary ammonium salt	upon treatment with strong base
Amide reduction	amides	
Nitrile reduction	nitriles	
Reduction of nitro compounds	nitro compounds	can be accomplished with elemental zinc, tin or iron with an acid.
Amine alkylation	haloalkane	
Delepine reaction	organohalide	reagent hexamine
Buchwald-Hartwig reaction	aryl halide	specific for aryl amines
Menshutkin reaction	tertiary amine	reaction product a quaternary ammonium cation
hydroamination	alkenes and alkynes	
Hofmann-Löffler reaction	haloamine	

Reductive Routes

Via the process of hydrogenation, nitriles are reduced to amines using hydrogen in the presence of a nickel catalyst. Reactions are sensitive acidic or alkaline conditions, which can cause hydrolysis of -CN group. $LiAlH_4$ is more commonly employed for the reduction of nitriles on the laboratory scale. Similarly, $LiAlH_4$ reduces amides to amines. Many amines are produced from aldehydes and ketones via reductive amination, which can either proceed catalytically or stoichiometrically. Aniline ($C_6H_5NH_2$) and its derivatives are prepared by reduction of the nitroaromatics. In industry, hydrogen is the preferred reductant, whereas in the laboratory, tin and iron are often employed.

Reactions

Alkylation, Acylation, and Sulfonation

Aside from their basicity, the dominant reactivity of amines is their nucleophilicity. Most primary amines are good ligands for metal ions to give coordination complexes. Amines are alkylated by alkyl halides. Acyl chlorides and acid anhydrides react with primary and secondary amines to form amides (the "Schotten-Baumann reaction").

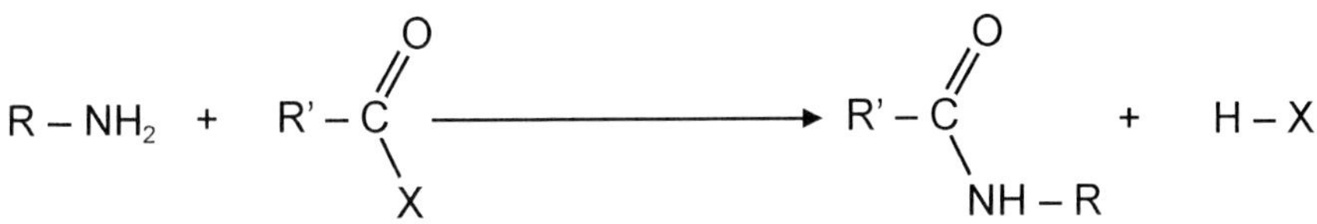

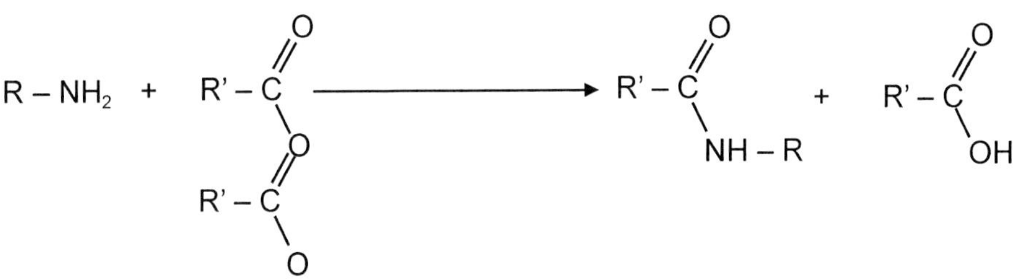

Similarly, with sulfonyl chlorides, one obtains sulfonamides. This transformation, known as the Hinsberg reaction, is a chemical test for the presence of amines. Because amines are basic, they neutralize acids to form the corresponding ammonium salts R_3NH^+. When formed from carboxylic acids and primary and secondary amines, these salts thermally dehydrate to form the corresponding amides.

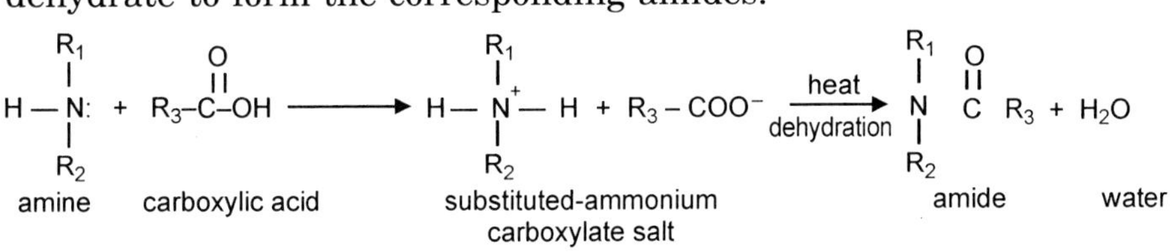

Diazotization

Amines react with nitrous acid to give diazonium salts. The alkyl diazonium salts are of little synthetic importance because they are too unstable. The most important members are derivatives of aromatic amines such as aniline ("phenylamine") (Ar = aryl or naphthyl):

$$ArNH_2 + HNO_2 + HX \rightarrow ArN_2^+X^- + 2\ H_2O$$

Anilines and naphthylamines form more stable diazonium salts, which can be isolated in the crystalline form. Diazonium salts undergo a variety of useful transformations involving replacement of the N_2 group with anions. For example, cuprous cyanide gives the corresponding nitriles:

$$ArN_2^+ + Y^- \rightarrow ArY + N_2$$

Aryldiazonium couple with electron-rich aromatic compounds such as a phenol to form azo compounds. Such reactions are widely applied to the production of dyes.

Conversion to Imines

Imine formation is an important reaction. Primary amines react with ketones and aldehydes to form imines. In the case of formaldehyde (R' = H), these products typically exist as cyclic trimers.

$$RNH_2 + R'_2C{=}O \rightarrow R'_2C{=}NR + H_2O$$

Reduction of these imines gives secondary amines:

$$R'_2C{=}NR + H_2 \rightarrow R'_2CH\text{-}NHR$$

Similarly, secondary amines react with ketones and aldehydes to form enamines:

$$R_2NH + R'(R''CH_2)C{=}O \rightarrow R''CH{=}C(NR_2)R' + H_2O$$

Biological Activity

Amines are ubiquitous in biology. The breakdown of amino acids releases amines, famously in the case of decaying fish which smell of trimethylamine. Many neurotransmitters are amines, including epinephrine, norepinephrine, dopamine, serotonin, and histamine. Protonated amino groups ($-NH_3^+$) are the most common positively charged moieties in proteins, specifically in the amino acid lysine. The anionic polymer DNA is typically bound to various amine-rich proteins. Additionally, the charged ammonium centers on lysine form salt bridges with carboxylate groups of other amino acids in polypeptides, which influences the three-dimensional structures of proteins.

Overview

An overview of the reactions of amine is given below:

Reaction name	Reaction product	Comment
Amine alkylation	amines	degree of substitution increases
Schotten-Baumann reaction	amides	Reagents: acyl chlorides, acid anhydrides
Hinsberg reaction	sulfonamides	Reagents: sulfonyl chlorides
Amine-carbonyl condensation	imines	
Organic oxidation	nitroso compounds	Reagent: peroxymonosulfuric acid
Organic oxidation	diazonium salt	Reagent: nitrous acid
Zincke reaction	Zincke aldehyde	reagent pyridinium salts, with primary and secondary amines
Emde degradation	tertiary amine	reduction of quaternary ammonium cations
Hofmann-Martius rearrangement	aryl substituted anilines	
Von Braun reaction	Organocyanamide	By cleavage (tertiary amines only) with cyanogen bromide
Hofmann Elimination	Alkene	proceedss by β-elimination of less hindered carbon
Cope reaction	Alkene	Similar to Hofmann elimination
carbylamine reaction	Isonitrile	(primary amines only)
Hoffmann's mustard oil test	Isothiocyanate	CS_2 and $HgCl_2$ are used. Thiocyanate smells like mustard.

Application of Amines—Dyes

Primary aromatic amines are used as a starting material for the manufacture of azo dyes. It reacts with nitric(III) acid to form diazonium salt, which can undergo coupling reaction to form azo compound. As azo-compounds are highly coloured, they are widely used in dyeing industries, such as:

- Methyl orange
- Direct brown 138
- Sunset yellow FCF
- Ponceau.

Drugs

Many drugs are designed to mimic or to interfere with the action of natural amine neurotransmitters, exemplified by the amine drugs:

- Chlorpheniramine is an antihistamine that helps to relieve allergic disorders due to cold, hay fever, itchy skin, insect bites and stings.
- Chlorpromazine is a tranquillizer that sedates without inducing sleep. It is used to relieve anxiety, excitement, restlessness or even mental disorder.
- Ephedrine and Phenylephrine, as amine hydrochlorides, are used as decongestants.
- Amphetamine, Methamphetamine, and Methcathinone are psychostimulant amines that are listed as controlled substances by the US DEA.
- Amitriptyline, Imipramine, Lofepramine and Clomipramine are tricyclic antidepressants and tertiary amines.
- Nortriptyline, Desipramine, and Amoxapine are tricyclic antidepressants and secondary amines. (The tricyclics are grouped by the nature of the final amine group on the side chain.)

Gas Treatment

Aqueous monoethanolamine (MEA), diglycolamine (DGA), diethanolamine (DEA), diisopropanolamine (DIPA) and methyldiethanolamine (MDEA) are widely used industrially for removing carbon dioxide (CO_2) and hydrogen sulfide (H_2S) from natural gas and refinery process streams. They may also be used to remove CO_2 from combustion gases/flue gases and may have potential for abatement of greenhouse gases. Related processes are known as sweetening.

Safety

Low molecular weight amines are toxic, and some are easily absorbed through the skin. Many higher molecular weight amines are highly active biologically.

Bibliography

Anderson, Irvine H.: *Aramco, the United States, and Saudi Arabia: A Study of the Dynamics of Foreign Oil Policy*, 1933-1950. Princeton: Princeton University Press, 1981.

Armistead, George: *Safety in Petroleum Refining and Related Industries*, New York: John G. Simmons & Co., 1950.

Balakrishan, A. : *Biofuel Crops and Renewable Energy*, Sublime Publications, Delhi, 2011.

Bhattacharya, K. G. and A. K. Talukdar: *Catalysis in Petroleum and Petrochemical Industries*, Alpha Science International, Delhi, 2005.

Black, Brian: *Petrolia: The Landscape of America's First Oil Boom.* Baltimore: Johns Hopkins University Press, 2000.

Blomen, Leo, and Michael Mugerwa: *Fuel Cell Systems.* New York: Plenum Press, 1993.

Bromley, Simon: *American Hegemony and World Oil: The Industry, the State System, and the World Economy.* University Park, Pa., 1991.

Burdick, Donald L., and William L. Leffler: *Petrochemicals in Nontechnical Language.* Tulsa, Okla.: PennWell, 2001.

Chester, Edward W.: *United States Oil Policy and Diplomacy: A Twentieth-Century Overview.* West-port, Conn.: Greenwood Press, 1983.

Chisholm, Archibald H. T.: *The First Kuwait Oil Concession Agreement: A Record of the Negotiations,* 1911-1934. London: Frank Cass, 1975.

Chowdhary, L.R. : *Petroleum Geology of the Cambay Basin, Gujarat, India*, Indian Petroleum, Delhi, 2004.

Clark, John G.: *The Political Economy of World Energy: A Twentieth-Century Perspective.* Chapel Hill, N.C., 1990.

Connelly, William L.: *The Oil Business As I Saw It: Half a Century with Sinclair.* Norman: University of Oklahoma Press, 1954.

Cordesman, Anthony H.: *The Gulf and the West: Strategic Relations and Military Realities.* Boulder, Colo.: Westview Press, 1988.

Cowhey, Peter F.: *The Problems of Plenty: Energy Policy and International Politics.* Berkeley: Uni-versity of California Press, 1985.

Cutler, Tom: *The Military Demand for Oil*. Petroleum Economist, London, 1989.

Davenport, E. H. and Sidney Russell Cooke: *The Oil Trusts and Anglo-American Relations*. Macmillan, 1924

DeLaisi, Francis: *Oil, Its Influence on Politics*. Labor Publishing,1923.

Diwan, Parag and Ashutosh Karnatak: *Handbook of Natural Gas Technology and Business*, Pentagon Energy Press, Delhi, 2009.

Eaton, Rev. S.J.M.: *Petroleum, a History of the Oil Region*, J.P. Skelly & Co., Philadelphia, 1866.

Engler, Robert: *The Politics of Oil: A Study of Private Power and Democratic Directions*. New York, Macmillan, 1961.

Feis, Herbert: *Petroleum and American Foreign Policy*. Stanford: Food Research Institute, 1944.

Fischer, Louis: *Oil Imperialism: The International Struggle for Petroleum*. New York: International Publishers, 1926.

Frankel, Paul: *Mattei: Oil and Power Politics*. New York and Washington: Praeger, 1966.

Garverick, Linda: *Corrosion in the Petrochemical Industry*. Materials Park, Ohio: ASM International, 2000.

Giddens, Paul H.: *The Birth of the Oil Industry*, The MacMillian Company, New York, 1938.

Gramling, Robert: *Oil on the Edge: Offshore Development, Conflict, Gridlock*. Albany: State University of New York Press, 1996.

Hamilton, Adrian: *Oil: The Price of Power*. London: Michael Joseph/Rainbird, 1986.

Hartshorn, J. E.: *Oil Companies and Governments: An Account of the International Oil Industry in Its Political Environment*. London: Faber and Faber, 1962.

Hyne, Norman J.: *Nontechnical Guide to Petroleum Geology, Exploration, Drilling, and Production*. PennWell Corporation, Delhi, 2001.

Jentleson, Bruce: *Pipeline Politics: The Complex Political Economy of East-West Energy Trade*. lthaca: Cornell University Press, 1986.

Kazemzadeh, Firuz: *The Struggle for Transcaucasia, 1917-1921*. Philosophical Library, 1951.

Kent, Marian: *Oil and Empire: British Policy and Mesopotamian Oil, 1900-1920*. London: Macmillan, 1976.

Klare, Michael: *Blood and Oil: The Dangers and Consequences of America's Growing Petroleum Diplomacy*. Metropolitan Books/ Henry Holt, 2004.

Klare, Michael: *Resource Wars: The New Landscape of Global Conflict*. Metropolitan Books / Henry Holt, 2001.

Kleveman, Lutz: *The New Great Game: Blood and Oil in Central Asia*. Atlantic Books, 2003.

Kordesch, Karl, and Günter Simader: *Fuel Cells and Their Applications*. New York: VCH, 1996.

Lieber, Robert J.: *Oil and the Middle East War: Europe in the Energy Crisis.* Cambridge: Center for International Affairs, 1976.

Lubell, Harold: *Middle East Oil Crises and Western Europe's Energy Supplies.* Baltimore: Johns Hopkins University Press, 1963.

Marsh, Steve: *Anglo-American Relations and Cold War Oil: Crisis in Iran.* New York: Palgrave MacMillan, 2003.

McLaurin, John J.: *Sketches in Crude-oil*, J. Horace McFarland Company, Harrisburg, 1896.

McNaugher, Thomas: *Arms and Oil: US Military Strategy and the Persian Gulf.* Brookings Institution, Londong, 1985.

McQuaig, Linda. *It's The Crude, Dude: War Big Oil, And The Fight For The Planet.* Doubleday Canada, 2004.

Mejcher, Helmut. *Imperial Quest for Oil: Iraq, 1910-1928.* Ithaca Press, NY, 1976.

Meyer, Lorenzo: *Mexico and the United States in the Oil Controversy: 1917-1942.* University of Texas Press, New York. 1977.

Miller, Aaron David: *Search for Security: Saudi Arabian Oil and American Foreign Policy, 1939-1949.* Chapel Hill: University of North Carolina Press, 1080.

Mohr, Anton. *The Oil War.* Harcourt, Brace, 1926. Hyperion Press, 1976.

Mosley, Leonard: *Power Play: Oil in the Middle East.* New York: Random House, 1973.

Naderpour, N and F Hematti Kordmahaleh: *Petroleum Geosciences*, SBS Pub, Delhi, 2009.

O'Connor, Harvey *The Empire of Oil.* Monthly Review Press, 1955

Odell, Peter R.: *Oil and World Power: Background of the Oil Crisis,* New York: Viking Penguin, 1986.

Olien, Diana Davids: *Oil in Texas: The Gusher Age, 1895–1945.* Austin: University of Texas Press, 2002.

Olien, Roger: *Oil and Ideology: The Cultural Creation of the American Petroleum Industry.* Chapel Hill: University of North Carolina Press, 2000.

Penrose, Edith, and E. F. Penrose: *Iraq: International Relations and National Development.* London: Ernest Bern, 1978.

Philip, George: *Oil and Politics in Latin America: Nationalist Movements and State Companies.* Cambridge University Press, Cambridge, 1982.

Randall, Stephen J. *United States Foreign Oil Policy, 1919–1948: For Profits and Security.* Kingston, Ontario, 1985.

Rathi, Rakesh : *Petroleum Refining Processes*, SBS Pub, Delhi, 2007.

Redlinger, Robert Y.: *Wind Energy in the 21st Century: Economics, Policy, Technology, and the Changing Electricity Industry.* New York: Macmillan, 2001.

Rister, Carl Coke: *Oil! Titan of the Southwest.* Norman: University of Oklahoma Press, 1949.

Sah, S.L. : *Encyclopaedia of Petroleum Science and Engineering*, Kalpaz, Delhi, 2003.

Sampson, Anthony: *The Seven Sisters: The Great Oil Companies and the World They Shaped.* New York, 1975.

Satyanarayana, D. : *Petroleum Geochemistry*, Daya Publishing, Delhi, 2011.

Schneider, Steven A.: *The Oil Price Revolution.* Baltimore, 1983.

Simmons, Matthew R.: *Twlight in the Desert: The Coming Saudi Oil Shock and the World Economy.* 2005

Singh, Lakshman and Anoop Tripathi: *Indian Petroleum Directory : With an Introductory Chapter on Petroleum Industry in India*, Indian Petroleum Pub, Delhi, 2007.

Speight, James G.: *The Chemistry and Terminology of Petroleum*. New York: Marcel Dekker, 1980.

Srivastava, Rajesh K. , Ch. Sivaji and N.V. Chalapathi Rao: *Indian Dykes: Geochemistry Geophysics and Geochronology*, Narosa, Delhi, 2008.

Stoff, Michael B. *Oil, War, and American Security*. Yale University Press, 1980.

Subramaniam, A P and S Balakrishna: *Advancing Frontiers in Geology and Geophysics,* Today and Tomorrow's, Delhi, 2007.

Tiwari, S.K. : *Fuel Minerals and Other Energy Resources*, Atlantic Pub, Delhi, 2010.

Tribedi, S. : *Proceedings of National Convention of Environment in India, Challenge for the 21st Century*, India, 1994.

Venn, Fiona: *Oil Diplomacy in the Twentieth Century.* New York, 1986. Pioneering study stressing oil's strategic importance.

Vervalin, Charles H.: *Fire Protection Manual for Hydrocarbon Processing Plants*. Houston: Gulf Publishing Co, 1985.

Vijayalakshmi, G S : *Fuels and Biofuels*, Agrobios, Delhi, 2007.

Watkins, Eric: *The Middle Eastern Environment*, London, The British Society for Middle Eastern Studie, 1995.

Wendel, Charles H.: *The Allis-Chalmers Story*. Sarasota, Florida: Crestline Publishing Co., 1988.

Yergin, Daniel: *The Prize: The Epic Quest for Oil, Money, and Power.* New York: Simon and Schuster, 1991.

Young, George J.: *Fuel Cells,* New York: Reinhold Publishing Corp., 1961.

Index

A

B

C

D

T

V

W

□□□